NAMES IN CURRENT USE IN THE FAMILIES TRICHOCOMACEAE, CLADONIACEAE, PINACEAE, AND LEMNACEAE

Regnum Vegetabile

a series of publications for the use of plant taxonomists published under
the auspices of the International Association for Plant Taxonomy
edited by Werner Greuter

Volume 128

ISSN 0080-0694

NCU – 2

Names in Current Use in the Families Trichocomaceae, Cladoniaceae, Pinaceae, and Lemnaceae

Edited by

Werner Greuter

on behalf of the
Special Committee on Names in Current Use

compiled by

J. I. Pitt, R. A. Samson, T. Ahti, A. Farjon

and

E. Landolt

1993
Published for the
International Association for Plant Taxonomy
by
Koeltz Scientific Books
D-61453 Königstein, Germany

ISBN 3-87429-348-3 (Germany)
ISBN 1-878762-44-3 (U.S.A.)

INTRODUCTION

By Werner Greuter

This is the second of a set of three *Regnum vegetabile* issues published under the auspices of the Special Committee on Names in Current Use. All three have served, among other things, to test the feasibility of such lists of names at various levels of the taxonomic hierarchy: the first (Regnum Veg. 126) for the rank of family; the present one for the infrageneric ranks, species in particular; and the next to follow for the rank of genus.

Whereas the two other items cover large portions of the plant kingdom(s), the present one must by necessity limit itself to a few selected examples. Four small to medium-sized families have been chosen, to stand for the huge diversity of organisms treated as plants and for the variety of problems affecting their nomenclature.

– The fungal family *Trichocomaceae* comprises organisms that are mainly studied and described on the basis of live material cultured in the laboratory. Type specimens therefore mostly consist of dried agar cultures, usually with parts of the same strain being kept alive for further examination. The family provides classical examples of species with a pleomorphic life cycle, in which an asexually propagating morph may alternate with a widely different, sexually reproducing one. By means of a special, fairly complex Article (Art. 59) the *Code* provides for an independent nomenclature of the anamorphs (asexual states) alone, of which the moulds of the genera *Aspergillus* and *Penicillium* are well-known, economically important representatives.

– The *Cladoniaceae* are lichenized fungi, i.e., symbiotic organisms which for nomenclatural purposes are reduced to their fungal component. They include dominant members of arctic and high-mountain plant communities such as the reindeer lichens, and some of their species have potential as bioindicators.

– Among the *Pinaceae* are some of the world's foremost timber trees and many other species that are important on a somewhat smaller scale in forestry and/or as ornamentals. They can aptly exemplify the bibliographic and nomenclatural difficulties found in plant groups which were studied traditionally, at least in part, by amateur botanists. They are the only one of the families treated in which hybrids have been described and named, the only one with a fossil record and with many described cultigens (fossils and cultigens were however excluded from coverage by the present list).

– The *Lemnaceae* are by far the smallest of the four groups. The duckweed family consists of minute, highly specialized aquatic angiosperms, again with a potential as bioindicators. They are the one case in which a modern, exhaustive monographic treatment is available to be followed.

Numerical analysis

Some figures will give an impression of the size and structure of each of the four individual lists.

Numbers of entries. – In the *Trichocomaceae* 614 binomials are listed under 27 genera, most of which (461, of 8 genera) pertain to asexual morphs and only 153 (but 19 genera) to sexually reproducing taxa. All but 17 of the latter have a corresponding named anamorph listed, but for a vast majority (325) of the named asexual taxa no sexually reproducing morph has so far been found and described. In the *Cladoniaceae* 11 genera are recognized, under which not only 422 species binomials but also 1 subgeneric, 9 sectional, 26 subspecific and 30 varietal names are listed. The *Pinaceae* comprise 12 genera; listed under these are 55 names of subdivisions of genera (mostly sections and subsections), 309 species binomials, and 194 infraspecific names (45 of subspecies, 149 of varieties). Last, the 4 Lemnaceous genera comprise 34 named species.

Alternative taxonomies. – Allowing for diverging taxonomic opinions is an important challenge faced by all compilers of NCU lists. How best to meet it must of course depend on the group treated and the number of ranks considered.

When only species names are listed, as in *Lemnaceae* and *Trichocomaceae,* the answer is simple. By adopting the narrowest species notion held by contemporary workers one will accommodate the nomenclatural needs of the "lumpers" and "splitters" alike. The question may still arise of how to list species assigned to different genera by different authors, but in both families generic limits now appear to be undisputed – which in the case of *Trichocomaceae* may reflect a trend toward consensus taxonomy monitored by an international group of appointed experts.

The two other lists have names at several ranks. They express taxonomic divergence not only by adopting a splitter's view in taxon circumscription but by listing side by side, on equal footing, synonyms based on the same type and available for use as alternatives. Synonymous entries can be recognized at having a cross-reference instead of a basionym and type mention. There are 66 of them in the *Cladoniaceae* list and no less than 110 under *Pinaceae.* Some reflect horizontal transfers (treatment of the same taxon, without change of rank, under different genera, or sometimes species), mostly in the lichen list where name pairs under *Cladina* and *Cladonia* account for 40 synonymies. Vertical transfer is particularly prominent in *Pinaceae* where no less than 103 rank changes between species, subspecies or variety and a further 3 between subgenus, section or subsection are expressed. Only exceptionally has a combined horizontal and vertical transfer been given recognition in the present lists.

Typification. – The proportion of definitely typified names is highest in *Trichocomaceae* (all but one out of a total of 614) and lowest in *Pinaceae* (266 of 448, or slightly less than 60 %). Four names in *Lemnaceae* (out of 34), and 11 (of 422) in *Cladoniaceae* also await type designation. (Names of generic subdivisions have been counted as belonging to the typified category even when their type is listed only as a binomial which has not itself a type specimen cited.) Holotypified names account for 38 % of the total in *Lemnaceae,* 47 % in both *Pinaceae* and *Trichocomaceae,* and 51 % in *Cladoniaceae.*

A careful, sensible type selection for names without a holotype – most of the older names belong to this category – is perhaps the most crucial condition for nomenclatural security and stability. The high number of lecto- and neotypifications newly proposed here by the compilers of the lists, always after critical evaluation of the type material, is therefore a particularly important contribution. There are 72 of them in the *Trichocomaceae* list, 109 in *Cladoniaceae* and 39 in *Pinaceae,* which corresponds of 22 %, 55 % and 71 %, respectively, of all lecto- and neotypes designated so far for the listed names. Irrespective of the fate of associated nomenclatural proposals, this feature alone will confer permanent value to the lists.

The actual and potential status of NCU lists

At present, the lists that follow are no more and no less than inventories of names used in current literature. They include thoroughly checked citations of their source and date of valid publication, basionym citations where appropriate, and indication of the nomenclatural type when possible. Such lists are useful by themselves, and will I am sure be welcomed by many. The consistent way in which the data are presented and citations are abbreviated may also be of some practical value as a model for nomenclatural purposes.

Some will be tempted to consider the listed names to be secure, rubber-stamped and vouched for, and it is indeed the declared wish of the editor and authors that they be such. It is however fair to warn that at present they are not, and that they will not be unless and until an International Botanical Congress gives them its special blessing. The nomenclatural status and the application of the names as listed is still open to controversy. In a few cases it is not even controversial but plainly untenable under the present nomenclatural rules.

There is one way to use the lists that should be explicitly discouraged. No one should waste his or her skill and efforts in trying to prove them faulty. They are deliberately wrong in some cases. Abandoning well established names on the grounds of nomenclatural technicalities does not further scientific knowledge, and worse, using them in an altered meaning is a major impediment to taxonomic progress. Therefore my plea: let sleeping dogs lie – until they die and be buried.

In my opinion there is a strong case for banning the uncertainty and abolishing the reward for futile nomenclatural digging exercises. The proposals made to the XV International Botanical Congress, to grant protected nomenclatural status to listed names in current use, would provide for the legal options that are needed, and under such options the present lists would be prime candidates for approval and consequent stabilization (of nomenclature, mind you, not taxonomy). In the first NCU list (Regnum Veg. 126) I have explained in some detail the principles and effects of nomenclatural protection. Here, rather than repeating myself, I will concentrate on aspects that are peculiar to the infrageneric ranks, where all names are combinations consisting of more than one element. The full proposals for NCU protection have been published in *Taxon* (40: 669-677. 1991).

Nomenclatural protection at ranks below genus

The proposals to provide for granting nomenclatural protection to listed names in current use had to be worded with particular care in order to be functional at the lower taxonomic ranks. It is not possible to protect epithets (they have no nomenclatural standing of their own), yet one would wish to maintain the epithet of a protected name when the taxon is transferred to another genus (or species) without change of rank. This is indeed what the NCU provisions are designed to achieve.

Under the proposed scheme, once a name is protected, all combinations with the same epithet and type, at the same rank, are also protected against unlisted competing synonyms. This holds irrespective of whether the combination one wants to adopt already exists or is yet to be validated. The only condition is that the epithet must be available in the required combination, i.e. that no tautonym would result and that no earlier homonym exists.

Just as priority, protection operates only at a given rank. There may be cases in which, under a given taxonomic concept, two or more protected names in the appropriate rank, while based on different types, are considered to be synonymous. In that case, the priority rules apply in establishing which of the protected names is correct, or which of their epithets must be adopted in the required combination. On the other hand, there may be taxa to which none of the protected names in the appropriate rank applies. Again, then, the normal priority rule will apply, but the choice will be among the unprotected names (if any) applying to the taxon.

These principles must be borne in mind when compiling a list of names for which nomenclatural protection is being sought, and when evaluating the usefulness of such a proposed list. When there is uncertainty over the adequate rank of a given taxon, it may be appropriate to list the name currently used for it in each of the ranks concerned. When the taxonomic position (generic affiliation) but not the rank is in doubt, it is much less important to list all currently used combinations explicitly, although this may still be useful for purposes such as the stabilization of authorship citation or date. Citation of basionyms of the listed names, or of the earliest combination in the appropriate rank, when different, will provide the date relevant for establishing priority. (By chance, in the context of the present lists no case has come to light in which the earliest combination in the appropriate rank is neither the listed name itself nor its basionym, but a third name.)

Authorship, date and place of valid publication

Same as for the family NCU lists, a neat distinction has been made between authors in the nomenclatural sense and authors of publications. The particle "in" and the personal names that may follow are not considered to be a part of the nomenclatural author citation but of the bibliographic reference. Such personal names are not therefore abbreviated but cited in full, in conformity with standard bibliographic procedure. Personal names that precede the "in" are, on the other hand, authors in the nomenclatural sense. The spelling of their names, their abbreviation if applicable, and the use of initials when appropriate, follow the standard of Brummitt & Powell's *Authors of plant names.* In those cases when the author(s) of the name and the author(s) of the validating publication are exactly the same they are, naturally, cited only once, and the nomenclatural citation standard is then followed; this is the situation when the particle "in" disappears and the author citation is followed immediately by a comma.

Joint authorship is indicated by the ampersand (&) that stands for Latin "et" (and); of three or more co-authors only the first is named, followed by "& al." Names of authors to whom a name has been ascribed by the real validating author(s) are mentioned ahead of the latter's name(s) from which they are separated by the particle "ex", it being understood that citation of this particle and the preceding name(s) it is optional. Names of pre-starting-point authors have, in principle, been omitted.

Citation of nomenclatural source, basionym (when applicable) and type (when known) are provided for all names recognized in the present booklet except for the generic ones where such data are to be included in the forthcoming list of generic names in current use (Regnum Veg. 129).

Great care has been devoted to the consistent citation of the nomenclatural sources of the listed names and their basionyms. All such sources were of course verified by the authors of the lists, so that the references given can be taken to be accurate and fulfilment of the conditions of valid publication can be vouched for. Bibliographic citations have been abbreviated in conformity with internationally recognized standards, as explained in the introduction to the previous lists (Regnum Veg. 126: 11) where further details on the adopted style of citation can be found. Here again, all journal and anonymous serial titles are preceded by the particle "in" but the titles of independently authored publications by a comma. Within each bibliographic citation, a colon (:) separates the elements that refer to the whole publication (title, edition, series, volume and fascicle number, as appropriate) from those relevant for the individual name (page and/or plate); and a period (.) separates the latter from the date of publication.

As a matter of principle a single source is considered to be relevant for the validation of a name, but for clarity and convenience a second, bracketed reference (an alternative title or a secondary, later source) has sometimes been added between square brackets.

The dates of publication have been given as exactly as possible. Full use has been made of the information included in *Taxonomic literature*, ed. 2, and in other scattered sources. Printed dates appearing in the publication itself have been accepted as correct unless they are demonstrably erroneous. The search for datings has not been exhaustive, but the cited dates should be amply sufficient for establishing the relative priority of potentially competing listed names.

Protection is being sought for the authorships, dates and places of valid publication of the listed names and basionyms. They have now been thoroughly verified, and it would seem a good idea to relieve future workers from the moral obligation of redoing the same work again, with a rate of "success" (or rather mischief) that I daresay would be minimal. Granting this kind of protection, same as protecting the spellings adopted here, would be an admittedly minor contribution to nomenclatural stability (although the dates of publication do matter when priority is concerned) but a welcome relief for practising taxonomists who judge that working with plants is scientifically more important and intellectually more rewarding than rechecking the bibliographic work of others.

Nomenclatural types

The nature of type specimens and the way in which they are cited may differ widely in different plant groups. Students of fungal cultures have developed good numbering systems but make little use of geographical provenance and original collection data, which are on the contrary paramount when referring to specimens of higher plants. It would have served no useful purpose to try and uniformize the way in which type specimens are cited in the present lists. Some comments on the subject may be found in the explanatory matter heading each list individually.

The question of whether it is appropriate to grant protected status to listed types, under the proposed NCU provisions, is particularly delicate and perhaps controversial. On one hand, it is certain that nomenclatural stabilization cannot in the long term dispense with protecting the standards governing the application of the names, i.e. the types. On the other hand one has to admit that being too rash in this respect may cause the sanctioning of errors that will be hard to correct, so that the overall benefit of protection might be greatly reduced.

The special and permanent nomenclature committees concerned will be asked for an opinion, to guide and advise the Nomenclature Section of the forthcoming International Botanical Congress in its actions. They will have to consider the type question with particular care. As pointed out previously, the extent to which typifications have been made and verified is not the same for all groups here treated. The coverage is virtually complete in the *Trichocomaceae* and *Cladoniaceae,* the two groups in which, as explained hereunder by the respective compilers, protecting the listed types would be particularly essential for nomenclatural stability. Postponing action on the types listed in the two other families might be less disturbing and could perhaps encourage additional type designations to be made in the six years to come.

Acknowledgements

The lists comprised in this booklet owe their substance and quality to the efforts of their compilers. They all have been extremely helpful and cooperative, and I have greatly appreciated their lenient attitude when a stringent editorial standardization and uniformization policy has led to reshufflings of format that they may not all have liked. It was indeed hard labour to edit a book of this kind, but thanks to their friendliness and patience it was also a pleasant experience. I want to thank them all once more, in public, and tell them how much I have been impressed by their energy, skill and experience.

Drafts of the lists have been widely circulated among major botanical institutions and different sets of specialists. The feedback has been substantial both in diversity and quality, as is acknowledged in the explanatory texts heading the individual lists. My personal thanks are, in addition, due to two most helpful persons: Paul Kirk of the C.A.B International Mycological Institute in Egham, U.K., who has checked several author names not to be found in Brummitt & Powell's standard book and has provided exact dates of publication for numerous journal issues cited in the *Trichocomaceae* list that were unavailable in Berlin; and Brigitte Zimmer, of the Berlin-Dahlem Museum, who ever so kindly and efficiently has once more taken care of the layout and produced the camera-ready copy for this book.

SPECIES NAMES IN CURRENT USE
IN THE
TRICHOCOMACEAE (FUNGI, EUROTIALES)

By J. I. Pitt and R. A. Samson

This compilation has been generated de novo on the basis of discussions between the authors and other members of the International Commission on *Penicillium* and *Aspergillus*. This Commission operates under the auspices of the Mycology Division of the International Union of Microbiological Societies (IUMS), a member union of the International Council of Scientific Unions (ICSU). Its present membership is as follows: Dr. J. I. Pitt, Australia; Dr. M. A. Klich, U.S.A.; Dr. R. A. Samson, Netherlands (Executive); Dr. R. H. Cruickshank, Australia; Dr. J. C. Frisvad, Denmark; Mrs. E. S. van Reenen-Hoekstra, Netherlands; Dr. Z. Lawrence, U.K.; Dr. E. J. Mullaney, USA; Dr. T. Okuda, Japan; Dr. K. A. Seifert, Canada; and Dr. S. S. Tzean, Republic of China. Draft lists were circulated among the membership of the Commission and sent to other mycologists who expressed interest in providing an input. In particular, valuable comments were received from Mrs. L. Schutte, South Africa, and Dr. S. Udagawa, Japan, which are gratefully acknowledged. Dr. H. C. Evans, International Institute for Biological Control, Ascot, Surrey, U.K., kindly provided valuable information on type specimens in the Kew Herbarium (K).

Very extensive discussions took place to ensure that the list finally generated represent the combined opinions of the experts involved. Important principles observed include: (1) where doubt exists about species circumscriptions, the splitters' view prevailed, to avoid the omission of "narrower" names required for use by some; (2) where a species is well defined, only a single name for each morph was accepted; and (3) varietal names were not considered, because in these genera conflicts over competing varieties within a single species have never occurred. Although the Commission's charter is specifically to study the systematics of just *Aspergillus* and *Penicillium*, it was felt logical to broaden the scope of this list to include the whole of the family *Trichocomaceae*.The circumscription of this family used here is that which was recently defined by Eriksson & Hawksworth (1990), except for *Chaetotheca*, which is of doubtful identity. Teleomorph genera included in this compilation are *Byssochlamys, Chaetosartorya, Cristaspora, Dendrosphaera, Dichlaena, Dichotomomyces, Emericella, Eupenicillium, Eurotium, Fennellia, Hemicarpenteles, Neosartorya, Penicilliopsis, Petromyces, Sclerocleista, Talaromyces, Thermoascus, Trichocoma* and *Warcupiella*. In addition, the associated anamorph genera *Aspergillus, Geosmithia, Merimbla, Paecilomyces, Penicillium* and *Torulomyces* have been included.

Several *Paecilomyces* species have teleomorphs that do not belong to the *Trichocomaceae* (Samson, 1974). However, for completeness all accepted *Paecilomyces* names are included in this list. Despite similarities to *Penicillium,* the genus *Nomuraea* has been excluded because its teleomorph affinities clearly lie with *Cordyceps.*

For lectotypes or neotypes the place where they were first designated is cited. The type specimens are cited in a much abbreviated form, mostly only by the number under which they are kept in the relevant reference collection. Only when available numbering is insufficient to unambiguously define the specimen are additional type data provided. Types referred to here are permanently preserved specimens, in line with the requirements of the current botanical *Code.* Some types referred to or designated here possess both teleomorph and anamorph states. The relevance of only a part of the type to a particular anamorph name is indicated in the text by the letters "p.p." (pro parte).

For valid publication after 1957, names of new taxa require indication of a permanently preserved type. It is unclear in several such cases whether authors of new species have indicated a permanent type, or merely a culture. Rather than conclude that publication of such names is invalid, which would at this time have serious repercussions on the stability of nomenclature for the whole family, we have taken the view that dried types may have existed, but are no longer extant. Herbarium types that have been designated by subsequent authors for such names (usually under the technically incorrect designation of "lectotypes"), or that are newly designated here, are therefore technically neotypes, and are so termed.

The collection in which each type is deposited is cited by the parenthetical herbarium abbreviation ("acronym") as given in Holmgren & al. (1990), except for special fungal collections not included in that work for which the abbreviations codified in the *World directory of collections of cultures and microorganisms* (Takishima & al., 1989) are used. The additional abbreviations and their explanation are as follows:

BKM, Institute of Biochemistry, Academy of Sciences, Puschino 142292, Russia;

CBM, Natural History Museum and Institute, Chiba 280, Japan;

FRR, C.S.I.R.O. Division of Food Science, North Ryde 2113, Australia;

IFM, Japan Federation of Culture Collections, Chiba 280, Japan;

NHL, National Institute of Hygienic Sciences, Tokyo 158, Japan;

PPEH, Department of Plant Pathology and Entomology, National Taiwan University, Taipei, Republic of China.

Where the anamorph of a listed holomorph species is known, its name has been added in square brackets, for the user's convenience. Note, however, that the anamorph names do not compete for priority with names based on a teleomorphic type.

Aspergillus Fr. : Fr.
(anamorphs)

Aspergillus acanthosporus Udagawa & Takada in Bull. Natl. Sci. Mus. Tokyo, ser. 2, 14: 503. 30 Sep 1971. – Holotype: No. 22462 p.p. (NHL)

Aspergillus aeneus Sappa in Allionia 2: 84. 1954. – Neotype (Samson & Gams, 1985): No. 128.58 (CBS).

Aspergillus albertensis J. P. Tewari in Mycologia 77: 114. 15 Feb 1985. – Holotype: No. 2976 p.p. (UAMH)

Aspergillus allahabadii B. S. Mehrotra & Agnihotri in Mycologia 54: 400. 7 Jan 1963. – Neotype (Samson & Gams, 1985): No. 164.63 (CBS).

Aspergillus alliaceus Thom & Church, Aspergilli: 163. Mar 1926. – Neotype (Malloch & Cain, 1972): No. 46232 p.p. (TRTC).

Aspergillus amazonensis (Henn.) Samson & Seifert in Samson & Pitt, Adv. Penicillium Aspergillus Syst.: 418. 1985. – Basionym: *Stilbothamnium amazonens* Henn. in Hedwigia 43: 396. 3 Sep 1904. – Holotype: Brazil, Jurna, Jul 1907, *Ule* in herb. Hennings (S)

Aspergillus ambiguus Sappa in Allionia 2: 254. 1955. – Neotype (Samson & Gams, 1985): No. 117.58 (CBS).

Aspergillus amylovorus Panas. ex Samson in Stud. Mycol. 18: 28. 3 Jan 1979. – Holotype: No. 600.67 (CBS).

Aspergillus anthodesmis Bartoli & Maggi in Trans. Brit. Mycol. Soc. 71: 386. 11 Jan 1979. – Holotype: No. 103 S (RO).

Aspergillus appendiculatus Blaser in Sydowia 28: 38. Dec 1975. – Holotype: No. 8286 (ZT).

Aspergillus arenarius Raper & Fennell, Genus Aspergillus: 475. 1965. – Neotype (Samson & Gams, 1985): No. 55632 (IMI).

Aspergillus asperescens Stolk in Antonie van Leeuwenhoek J. Microbiol. Serol. 20: 303. 1954. – Neotype (designated here): No. 46813 (IMI).

Aspergillus atheciellus Samson & W. Gams in Samson & Pitt, Adv. Penicillium Aspergillus Syst.: 34. 1985. – Holotype: No. 32048 p.p. (IMI).

Aspergillus aureolatus Munt.-Cvetk. & Bata in Bull. Inst. Jard. Bot. Univ. Beograd, ser. 2, 1: 196. 1964. – Neotype (Samson & Gams, 1985): No. 190.65 (CBS).

Aspergillus aureoluteus Samson & W. Gams in Samson & Pitt, Adv. Penicillium Aspergillus Syst.: 34. 1985. – Holotype: No. 105.55 p.p. (CBS).

Aspergillus auricomus (Guég.) Saito in J. Ferment. Technol. 17: 3. 1939. – Basionym: *Sterigmatocystis auricoma* Guég. in Bull. Soc. Mycol. France 15: 171. 31 Jul 1899. – Neotype (Samson & Gams, 1985): No. 467.65 (CBS).

Aspergillus avenaceus G. Sm. in Trans. Brit. Mycol. Soc. 26: 24. 8

Apr 1943. – Neotype (Samson & Gams, 1985): No. 109.46 (CBS).

Aspergillus awamorii Nakaz. in Rep. Gov. Res. Inst. Formosa 1907: 1. 1907. – Neotype (Al-Musallam, 1980): No. 557.65 (CBS).

Aspergillus bicolor M. Chr. & States in Mycologia 70: 337. 25 Mai 1978. – Holotype: No. RMF 2058 p.p. (NY).

Aspergillus biplanus Raper & Fennell, Genus Aspergillus: 434. 1965. – Neotype (Samson & Gams, 1985): No. 235602 (IMI).

Aspergillus bisporus Kwon-Chung & Fennell in Mycologia 63: 479. 4 Jun 1971. – Holotype: NRRL 3693 p.p. (BPI).

Aspergillus brevipes G. Sm. in Trans. Brit. Mycol. Soc. 35: 241. 19 Dec 1952. – Holotype: [dried culture from soil from] Australia, Mt. Kosciusko, *s.coll.* (K).

Aspergillus bridgeri M. Chr. in Mycologia 74: 210. 2 Apr 1982. – Holotype: No. JB 26-1-2 (NY).

Aspergillus brunneouniseriatus Suj. Singh & B. K. Bakshi in Trans. Brit. Mycol. Soc. 44: 160. 20 Jun 1961. – Neotype (Samson & Gams, 1985): No. 227677 (IMI).

Aspergillus brunneus Delacr. in Bull. Soc. Mycol. France 9: 185. 1893. – Neotype (Samson & Gams, 1985): No. 211378 p.p. (IMI).

Aspergillus caesiellus Saito in J. Coll. Sci. Imp. Univ. Tokyo 18(5): 49. 28 Jan 1904. – Neotype (Samson & Gams, 1985): No. 172278 (IMI).

Aspergillus caespitosus Raper & Thom in Mycologia 36: 563. Nov-Dec 1944. – Neotype (Samson & Gams, 1985): No. 16034ii (IMI).

Aspergillus campestris M. Chr. in Mycologia 74: 212. 2 Apr 1982. – Holotype: No. ST 2-3-1 (NY).

Aspergillus candidus Link in Ges. Naturf. Freunde Berlin Mag. Neuesten Entdeck. Gesammten Naturk. 3: 16. 1809 : Fr. – Neotype (Samson & Gams, 1985): No. 567.65 (CBS).

Aspergillus carbonarius (Bainier) Thom in J. Agric. Res. 7: 12. 2 Oct 1916. – Basionym: *Sterigmatocystis carbonaria* Bainier in Bull. Soc. Bot. France 27: 27. 1880. – Neotype (Al-Musallam, 1980): No. 556.65 (CBS).

Aspergillus carneus Blochwitz in Ann. Mycol. 31: 81. 25 Jan 1933. – Neotype (Samson & Gams, 1985): No. 1358818 (IMI).

Aspergillus cervinus Massee in Bull. Misc. Inform. Kew 1914: 158. 10 Jun 1914. – Neotype (Christensen & Fennell, 1964): No. WISC WT 540 (WIS).

Aspergillus chevalieri L. Mangin in Ann. Sci. Nat., Bot., ser. 9, 10: 362. 1909. – Neotype (Kozakiewicz, 1989): No. 211382 p.p. (IMI).

Aspergillus chryseides Samson & W. Gams in Samson & Pitt, Adv. Penicillium Aspergillus Syst.: 36. 1985. – Holotype: No. 238612 (IMI).

Aspergillus citrisporus Höhn. in Sitzungsber. Kaiserl. Akad. Wiss., Math.-Naturwiss. Cl., Abt. 1, 111: 1036. 1902. – Holotype (Subramanian, 1972): ex caterpillar dung, Kittery Point, *R. Thaxter* (FH).

Aspergillus clavatoflavus Raper & Fennell, Genus Aspergillus: 378. 1965. – Neotype (Samson & Gams, 1985): No. 124937 (IMI).

Aspergillus clavatonanicus Bat. & al. in Anais Fac. Med. Univ. Recife 15: 197. 1955. – Neotype (Samson & Gams, 1985): No. 235352 (IMI).

Aspergillus clavatus Desm. in Ann. Sci. Nat., Bot., ser. 2, 2: 71. 1834. – Neotype (Samson & Gams, 1985): No. 15949 (IMI).

Aspergillus compatibilis Samson & W. Gams in Samson & Pitt, Adv. Penicillium Aspergillus Syst.: 42. 1985. – Holotype: No. 488.65 (CBS).

Aspergillus conicus Blochwitz in Ann. Mycol. 12: 38. 10 Feb 1914. – Neotype (Samson & Gams, 1985): No. 172281 (IMI).

Aspergillus conjunctus Kwon-Chung & Fennell in Raper & Fennell, Genus Aspergillus: 552. 1965. – Neotype (Samson & Gams, 1985): No. 135421 (IMI).

Aspergillus coremiiformis Bartoli & Maggi in Trans. Brit. Mycol. Soc. 71: 386. 11 Jan 1979. – Holotype: No. 102 S (RO).

Aspergillus corrugatus Udagawa & Y. Horie in Mycotaxon 4: 535. 18 Dec 1976. – Holotype: No. 2763 p.p. (NHL).

Aspergillus cremeoflavus Samson & W. Gams in Samson & Pitt, Adv. Penicillium Aspergillus Syst.: 37. 1985. – Holotype: No. 123749ii (IMI).

Aspergillus cristatellus Kozak. in Mycol. Pap. 161: 81. 30 Jun 1989. – Holotype: No. 172280 p.p. (IMI).

Aspergillus crustosus Raper & Fennell, Genus Aspergillus: 532. 1965. – Neotype (Samson & Gams, 1985): No. 135819 (IMI).

Aspergillus curviformis H. J. Chowdhery & J. N. Rai in Nova Hedwigia 32: 231. 11 Feb 1980. – Holotype: [dried culture from soil from] India, Kagh Islands, *s.coll.* (IWG).

Aspergillus crystallinus Kwon-Chung & Fennell in Raper & Fennell, Genus Aspergillus: 471. 1965. – Neotype (Samson & Gams, 1985): No. 139270 (IMI).

Aspergillus deflectus Fennell & Raper in Mycologia 47: 83. 8 Mar 1955. – Neotype (Samson & Gams, 1985): No. 61448 (IMI).

Aspergillus dimorphicus B. S. Mehrotra & Prasad in Trans. Brit. Mycol. Soc. 52: 331. 25 Apr 1969. – Neotype (Kozakiewicz, 1989): No. 131553 (IMI).

Aspergillus diversus Raper & Fennell, Genus Aspergillus: 437. 1965. – Neotype (Samson & Gams, 1985): No. 232882 (IMI).

Aspergillus duricaulis Raper & Fennell, Genus Aspergillus: 249.

1965. – Neotype (Samson & Gams, 1985): No. 172282 (IMI).

Aspergillus dybowskii (Pat.) Samson & Seifert in Samson & Pitt, Adv. Penicillium Aspergillus Syst.: 422. 1985. – Basionym: *Penicilliopsis dybowskii* Pat. in Bull. Soc. Mycol. France 7: 54. 1891. – Lectotype (designated here; see also Samson & Seifert, 1985): Congo, Jan 1894, *Dybowski* in herb. Bresadola (S).

Aspergillus eburneocremeus Sappa in Allionia 2: 87. 1954. – Neotype (Samson & Gams, 1985): No. 69856 (IMI).

Aspergillus egyptiacus Moub. & Mustafa in Egypt. J. Bot. 15: 153. 1972. – Neotype (Samson & Gams, 1985): No. 141415 (IMI).

Aspergillus elegans Gasperini in Atti Soc. Tosc. Sci. Nat. Pisa, Processi Verbali 8: 328. 1887. – Neotype (Samson & Gams, 1985): No. 102.14 (CBS).

Aspergillus ellipticus Raper & Fennell, Genus Aspergillus: 319. 1965. – Neotype (Al-Musallam, 1980): No. 707.79 (CBS).

Aspergillus elongatus J. N. Rai & S. C. Agarwal in Canad. J. Bot. 48: 791. 29 Mai 1970. – Neotype (Samson & Gams, 1985): No. 387.75 (CBS).

Aspergillus erythrocephalus Berk. & M. A. Curtis in J. Linn. Soc., Bot., 10: 362. 16 Jun 1868. – Holotype: Cuba, *Wright 764* (K).

Aspergillus falconensis Y. Horie in Trans. Mycol. Soc. Japan 30: 257.

Oct 1989. – Holotype: No. 10001 p.p. (CBM).

Aspergillus fennelliae Kwon-Chung & S. J. Kim in Mycologia 66: 629. 28 Aug 1974. – Neotype (Kozakiewicz, 1989): No. 278382 (IMI).

Aspergillus flaschentraegeri Stolk in Trans. Brit. Mycol. Soc. 47: 123. 10 Apr 1964. – Neotype (Samson & Gams, 1985): No. CBS 108.63 (K).

Aspergillus fischerianus Samson & W. Gams in Samson & Pitt, Adv. Penicillium Aspergillus Syst.: 39. 1985. – Holotype: No. 21139ii p.p. (IMI).

Aspergillus flavipes (Bainier & Sartory) Thom & Church, Aspergilli: 155. Mar 1926. – Basionym: *Sterigmatocystis flavipes* Bainier & Sartory in Bull. Soc. Mycol. France 27: 90. 1911. – Neotype (Samson & Gams, 1985): No. 171885 (IMI).

Aspergillus flavofurcatus Bat. & H. Maia in Anais Soc. Biol. Pernambuco 13: 94-96. 1955. – Neotype (Samson & Gams, 1985): No. 124938 (IMI).

Aspergillus flavus Link in Ges. Naturf. Freunde Berlin Mag. Neuesten Entdeck. Gesammten Naturk. 3: 16. 1809 : Fr. – Neotype (Kozakiewicz, 1982): No. 124930 (IMI).

Aspergillus floriformis Samson & Mouch. in Antonie van Leeuwenhoek J. Microbiol. Serol. 40: 343. 1975. – Holotype: No. 937.73 (CBS).

Aspergillus foeniculicola Udagawa in Trans. Mycol. Soc. Japan 20: 13. Jun 1979. – Holotype: No. 2777 p.p. (NHL).

Aspergillus foetidus Thom & Raper, Man. Aspergilli: 219. 1945. – Neotype (Samson & Gams, 1985): No. 121.28 (CBS).

Aspergillus foveolatus Y. Horie in Trans. Mycol. Soc. Japan 19: 313. Oct 1978. – Holotype: No. 4547 p.p. (IFM).

Aspergillus fruticans Samson & W. Gams in Samson & Pitt, Adv. Penicillium Aspergillus Syst.: 40. 1985. – Holotype: No. 139279 p.p. (IMI).

Aspergillus fumigatus Fresen., Beitr. Mykol.: 81. 18 Aug 1863. – Neotype (Samson & Gams, 1985): No. 16152 (IMI).

Aspergillus funiculosus G. Sm. in Trans. Brit. Mycol. Soc. 39: 111. 20 Mar 1956. – Neotype (Samson & Gams, 1985): No. 44397 (IMI).

Aspergillus giganteus Wehmer in Mém. Soc. Phys. Genève 33(2): 85. Aug 1901. – Neotype (Samson & Gams, 1985): No. 227678 (IMI).

Aspergillus glaber Blaser in Sydowia 28: 35. Dec 1975. – Holotype: No. 8218T p.p. (ZT).

Aspergillus glaucoaffinis Samson & W. Gams in Samson & Pitt, Adv. Penicillium Aspergillus Syst.: 47. 1985. – Holotype: No. 16122ii (IMI).

Aspergillus glauconiveus Samson & W. Gams in Samson & Pitt, Adv. Penicillium Aspergillus Syst.: 45. 1985. – Holotype: No. 32050ii (IMI).

Aspergillus glaucus Link, in Ges. Naturf. Freunde Berlin Mag. Neuesten Entdeck. Gesammten Naturk. 3: 16. 1809 : Fr. – Neotype (designated here): No. 211383 (IMI).

Aspergillus globosus H. J. Chowdhery & J. N. Rai in Nova Hedwigia 32: 233. 11 Feb 1980. – Holotype: [dried culture from soil from] India, Kagh Islands, *s.coll.* (LWG).

Aspergillus gorakhpurensis Kamal & Bhargava in Trans. Brit. Mycol. Soc. 52: 338. 25 Apr 1969. – Neotype (Samson & Gams, 1985): No. 130728 (IMI).

Aspergillus gracilis Bainier in Bull. Soc. Mycol. France 23: 90. 1907. – Neotype (Samson & Gams, 1985): No. 211393 (IMI).

Aspergillus granulosus Raper & Thom in Mycologia 36: 565. Nov-Dec 1944. – Neotype (Kozakiewicz, 1989; incorrectly listed as No. 172278 (IMI) by Samson & Gams, 1985): No. 17278ii (IMI).

Aspergillus halophilicus C. M. Chr. & al. in Mycologia 51: 636. 17 Mar 1961. – Neotype (Samson & Gams, 1985): No. NRRL 2739 p.p. (BPI).

Aspergillus helicothrix Al-Musallam in Antonie van Leeuwenhoek J. Microbiol. Serol. 46: 407. 1980. – Holotype: No. 677.79 (CBS).

Aspergillus heterocaryoticus C. M. Chr. & al. in Mycologia 57: 535. 2

Aug 1965. – Holotype: No. NCF C-100 p.p. (BPI).

Aspergillus heteromorphus Bat. & H. Maia in Anais Soc. Biol. Pernambuco 15: 200. 1957. – Neotype (Samson & Gams, 1985): No. 172288 (IMI).

Aspergillus hiratsukae Udagawa & al. in Trans. Mycol. Soc. Japan 32: 23. Apr 1991. – Holotype: No. 3008 p.p. (NHL).

Aspergillus igneus Kozak. in Mycol. Pap. 161: 52. 30 Jun 1989. – Holotype: No. 75886 (IMI).

Aspergillus insulicola Montem. & A. R. Santiago in Mycopathologia 55: 130. 30 Apr 1975. – Neotype (Samson & Gams, 1985): No. 382.75 (CBS).

Aspergillus intermedius Blaser in Sydowia 28: 41. Dec 1976. – Neotype (Kozakiewicz, 1989): No. 89278 p.p. (IMI).

Aspergillus itaconicus Kinosh. in Bot. Mag. (Tokyo) 45: 60. 1931. – Neotype (Samson & Gams, 1985): No. 16119 (IMI).

Aspergillus ivoriensis Bartoli & Maggi in Trans. Brit. Mycol. Soc. 71: 383. 11 Jan 1979. – Holotype: No. 101 S (RO).

Aspergillus janus Raper & Thom in Mycologia 36: 556. Nov-Dec 1944. – Neotype (Samson & Gams, 1985): No. 16065 (IMI).

Aspergillus japonicus Saito in Bot. Mag. (Tokyo) 20: 61. Jun 1906. – Neotype (Samson & Gams, 1985): No. 114.51 (CBS).

Aspergillus kanagawaënsis Nehira in J. Jap. Bot. 26: 109. Apr 1951. – Neotype (Samson & Gams, 1985): No. 126690 (IMI).

Aspergillus lanosus Kamal & Bhargava in Trans. Brit. Mycol. Soc. 52: 336. 25 Apr 1969. – Neotype (Samson & Gams, 1985): No. 130727 (IMI).

Aspergillus leporis States & M. Chr. in Mycologia 58: 738. 9 Nov 1966. – Holotype: No. RMF 99 (NY).

Aspergillus leucocarpus Hadlok & Stolk in Antonie van Leeuwenhoek J. Microbiol. Serol. 35: 9. 1969. – Holotype: No. 353.68 p.p. (CBS).

Aspergillus longivesica L. H. Huang & Raper in Mycologia 63: 53. 10 Mar 1971. – Holotype: No. N1129 (WIS).

Aspergillus lucknowensis J. N. Rai & al. in Canad. J. Bot. 46: 1483. 20 Dec 1968. – Neotype (designated here): No. 449.75 (CBS).

Aspergillus malodoratus Kwon-Chung & Fennell in Raper & Fennell, Genus Aspergillus: 468. 1965. – Neotype (Samson & Gams, 1985): No. 172289 (IMI).

Aspergillus maritimus Samson & W. Gams in Samson & Pitt, Adv. Penicillium Aspergillus Syst.: 43. 1985. – Holotype: India, Kagh Islands, *s.coll.* (LWG).

Aspergillus medius R. Meissn. in Bot. Zeitung, 2. Abt., 55: 356. 1 Dec 1897. – Neotype (Samson & Gams, 1985): No. 113.27 (CBS).

Aspergillus melleus Yukawa in J. Coll. Agric. Imp. Univ. Tokyo 1: 358. 28 Mar 1911. – Neotype (Samson & Gams, 1985): No. 546.65 (CBS).

Aspergillus microcysticus Sappa in Allionia 2: 251. 1955. – Neotype (Samson & Gams, 1985): No. 139275 (IMI).

Aspergillus microthecius Samson & W. Gams in Samson & Pitt, Adv. Penicillium Aspergillus Syst.: 46. 1985. – Holotype: No. 139280 p.p. (IMI).

Aspergillus multicolor Sappa in Allionia 2: 87. 1954. – Neotype (Samson & Gams, 1985): No. 69875 (IMI).

Aspergillus navahoënsis M. Chr. & States in Mycologia 74: 226. 2 Apr 1982. – Holotype: No. SD-5 p.p. (NY).

Aspergillus neocarnoyi Kozak. in Mycol. Pap. 161: 63. 30 Jun 1989. – Holotype: No. 172279 p.p. (IMI).

Aspergillus neoglaber Kozak. in Mycol. Pap. 161: 56. 30 Jun 1989. – Holotype: No. 61447 (IMI).

Aspergillus nidulans (Eidam) G. Winter in Rabenh. Krypt.-Fl., ed. 2, 1(2): 62. Mar 1884, *nom. cons. prop.* – Basionym: *Sterigmatocystis nidulans* Eidam in Beitr. Biol. Pflanzen 3: 392. 1883. – Neotype (Kozakiewicz & al., 1992): No. 86806 (IMI).

Aspergillus niger Tiegh. in Ann. Sci. Nat., Bot., ser. 5, 8: 240. Oct 1867, *nom. cons. prop.* – Neotype

(Al-Musallam, 1980): No. 554.65 (CBS).

Aspergillus niveus Blochwitz in Ann. Mycol. 27: 205. 25 Jun 1929. – Neotype (Samson & Gams, 1985): No. 171878 (IMI).

Aspergillus nomius Kurtzman & al. in Antonie van Leeuwenhoek J. Microbiol. Serol. 53: 151. 1987. – Holotype: No. NRRL 13137 (BPI).

Aspergillus nutans McLennan & Ducker in Austral. J. Bot. 2: 355. Nov 1954. – Neotype (Samson & Gams, 1985): No. 62874ii (IMI).

Aspergillus ochraceoroseus Bartoli & Maggi in Trans. Brit. Mycol. Soc. 71: 393. 11 Jan 1979. – Holotype: No. 104 S (RO).

Aspergillus ochraceus K. Wilh., Beitr. Kenntn. Aspergillus: 66. 28 Apr 1877. – Neotype (designated here): No. 16247iv (IMI).

Aspergillus ornatulus Samson & W. Gams in Samson & Pitt, Adv. Penicillium Aspergillus Syst.: 45. 1985. – Holotype: No. 55295 p.p. (IMI).

Aspergillus oryzae (Ahlb.) Cohn in Jahresber. Schles. Ges. Vaterl. Cult. 61: 226. 1884. – Basionym: *Eurotium oryzae* Ahlb. in Dingler's Polytechn. J. 230: 330. 1878. – Neotype (Samson & Gams, 1985): No. 16266 (IMI).

Aspergillus ostianus Wehmer in Bot. Centralbl. 80: 461. 13 Dec 1899. – Neotype (Samson & Gams, 1985): No. 15960 (IMI).

Aspergillus paleaceus Samson & W. Gams in Samson & Pitt, Adv.

Penicillium Aspergillus Syst.: 50. 1985. – Neotype (Samson & Gams, 1985): No. 172293 p.p. (IMI).

Aspergillus pallidus Kamyschko in Bot. Mater. Otd. Sporov. Rast. 16: 93. 19 Jun 1963. – Neotype (Samson & Gams, 1985): No. 129967 (IMI).

Aspergillus panamensis Raper & Thom in Mycologia 36: 568. Nov-Dec 1944. – Neotype (Samson & Gams, 1985): No. 19393iii (IMI).

Aspergillus paradoxus Fennell & Raper in Mycologia 47: 69. 8 Mar 1955. – Neotype (Samson & Gams, 1985): No. 117502 (IMI).

Aspergillus parasiticus Speare in Bull. Div. Pathol. Physiol., Hawaiian Sugar Planters' Assoc. Exp. Sta. 12: 38. 1912. – Neotype (Kozakiewicz, 1982): No. 15957ix (IMI).

Aspergillus parvulus G. Sm. in Trans. Brit. Mycol. Soc. 44: 45. 21 Mar 1961. – Holotype: No. 86558 (IMI).

Aspergillus penicillioides Speg. in Revista Fac. Agron. Univ. Nac. La Plata 2: 246. 1896. – Neotype (Samson & Gams, 1985): No. 211342 (IMI).

Aspergillus petrakii Vörös in Sydowia, ser. 2, Beih., 1: 62. Jan-Aug 1957. – Neotype (Samson & Gams, 1985): No. 172291 (IMI).

Aspergillus peyronelii Sappa in Allionia 2: 248. 1955. – Neotype (Samson & Gams, 1985): No. 139271 (IMI).

Aspergillus phoenicis (Corda) Thom. in J. Agric. Res. 7: 14. 2 Oct 1916. – Basionym: *Ustilago phoenicis* Corda, Icon. Fung. 4: 9. Sep 1840. – Holotype: [on dates from] Turkey, Istanbul, 1837, *Schmidt* (PRM).

Aspergillus protuberus Munt.-Cvetk. in Mikrobiologija 5: 119. 1968. – Neotype (Samson & Gams, 1985): No. 602.74 (CBS).

Aspergillus pseudodeflectus Samson & Mouch. in Antonie van Leeuwenhoek J. Microbiol. Serol. 40: 345. 1975. – Holotype: No. 756.74 (CBS).

Aspergillus pulverulentus (McAlpine) Wehmer in Centralbl. Bakteriol., 2. Abth., 18: 394. 22 Apr 1907. – Basionym: *Sterigmatocystis pulverulenta* McAlpine in Agric. Gaz. New South Wales 7: 302. 1897. – Holotype: [on *Phaseolus vulgaris* from] Australia, Victoria, Burnley Bot. Garden, *McAlpine* (VPRI).

Aspergillus pulvinus Kwon-Chung & Fennell in Raper & Fennell, Genus Aspergillus: 45. 1965. – Neotype (Samson & Gams, 1985): No. 139628 (IMI).

Aspergillus puniceus Kwon-Chung & Fennell in Raper & Fennell, Genus Aspergillus: 547. 1965. – Neotype (Samson & Gams, 1985): No. 126692 (IMI).

Aspergillus purpureus Samson & Mouch. in Antonie van Leeuwenhoek J. Microbiol. Serol. 41: 350. 1975. – Holotype: No. 754.74 p.p. (CBS).

Aspergillus quadricingens Kozak. in Mycol. Pap. 161: 54. 30 Jun 1989. – Holotype: No. 48583ii p.p. (IMI).

Aspergillus raperi Stolk in Trans. Brit. Mycol. Soc. 40: 190. 1 Jul 1957. – Holotype: [dried culture from soil from] Zaire, Yangambi, *Meyer* (K).

Aspergillus recurvatus Raper & Fennell, Genus Aspergillus: 529. 1965. – Neotype (Samson & Gams, 1985): No. 36528 (IMI).

Aspergillus reptans Samson & W. Gams in Samson & Pitt, Adv. Penicillium Aspergillus Syst.: 48. 1985 (≡ *Aspergillus glaucus* var. *repens* Corda, Icon. Fung. 5: 53. Jun 1842 [non *Aspergillus repens* (de Bary) E. Fisch. in Engler & Prantl, Nat. Pflanzenfam. 1(1): 302. Feb 1897]). – Neotype (Samson & Gams, 1985): No. 529.65 p.p. (CBS).

Aspergillus restrictus G. Sm. in J. Textile Inst. 22: 115. 1931. – Neotype (Samson & Gams, 1985): No. 16267 (IMI).

Aspergillus robustus M. Chr. & Raper in Mycologia 70: 200. 27 Mar 1978. – Holotype: No. WB 5286 (NY).

Aspergillus rubrobrunneus Samson & W. Gams in Samson & Pitt, Adv. Penicillium Aspergillus Syst.: 49. 1985. – Holotype: No. 530.65 p.p. (CBS).

Aspergillus rugulovalvus Samson & W. Gams in Samson & Pitt, Adv. Penicillium Aspergillus

Syst.: 49. 1985. – Holotype: No. 136775 p.p. (IMI).

Aspergillus sclerotiorum G. A. Huber in Phytopathology 23: 306. Mar 1933. – Neotype (Samson & Gams, 1985): No. 56673 (IMI).

Aspergillus sepultus Tuthill & M. Chr. in Mycologia 78: 475. 16 Jun 1986. – Holotype: No. RMF 7602 (NY).

Aspergillus silvaticus Fennell & Raper in Mycologia 47: 83. 8 Mar 1955. – Neotype (Samson & Gams, 1985): No. 61456 (IMI).

Aspergillus sojae Sakag. & K. Yamada ex Murak. in Rep. Res. Inst. Brewing 143: 8. 1971. – Neotype (designated here): No. 191300 (IMI).

Aspergillus sparsus Raper & Thom in Mycologia 36: 572. Nov-Dec 1944. – Neotype (Samson & Gams, 1985): No. 19394 (IMI).

Aspergillus spathulatus Takada & Udagawa in Mycotaxon 24: 396. 11 Dec 1985. – Holotype: No. 2947 p.p. (NHL).

Aspergillus spectabilis M. Chr. & Raper in Mycologia 70: 333. 25 Mai 1978. – Holotype: No. RMFH 429 p.p. (NY).

Aspergillus spelunceus Raper & Fennell, Genus Aspergillus: 457. 1965. – Neotype (Samson & Gams, 1985): No. 211389 (IMI).

Aspergillus spinosus Kozak. in Mycol. Pap. 161: 58. 30 Jun 1989. – Holotype: No. 211390 (IMI).

Aspergillus stellifer Samson & W. Gams in Samson & Pitt, Adv.

Penicillium Aspergillus Syst.: 52. 1985. – Holotype: Bowenpilly near Secundarabad, *s.coll.*, (K).

Aspergillus striatulus Samson & W. Gams in Samson & Pitt, Adv. Penicillium Aspergillus Syst.: 50. 1985. – Holotype: No. 96679 (IMI).

Aspergillus stromatoides Raper & Fennell, Genus Aspergillus: 421. 1965. – Neotype (Samson & Gams, 1985): No. 123750 (IMI).

Aspergillus sublatus Y. Horie in Trans. Mycol. Soc. Japan 20: 481. Dec 1979. – Holotype: No. 4553 p.p. (IFM).

Aspergillus subolivaceus Raper & Fennell, Genus Aspergillus: 385. 1965. – Neotype (Samson & Gams, 1985): No. 44882 (IMI).

Aspergillus subsessilis Raper & Fennell, Genus Aspergillus: 530. 1965. – Neotype (Samson & Gams, 1985): No. 135820 (IMI).

Aspergillus sulphureus (Fresen.) Wehmer in Mém. Soc. Phys. Genève 33(2): 113. Aug 1901. – Basionym: *Sterigmatocystis sulphurea* Fresen., Beitr. Mykol.: 83. 18 Aug 1863. – Neotype (Kozakiewicz, 1989): No. 211397 (IMI).

Aspergillus sydowii (Bainier & Sartory) Thom & Church, Aspergilli: 147. Mar 1926. – Basionym: *Sterigmatocystis sydowii* Bainier & Sartory in Ann. Mycol. 11: 25. 15 Mar 1913. – Neotype (Samson & Gams, 1985): No. 211384 (IMI).

Aspergillus tamarii Kita in Centralbl. Bakteriol., 2. Abth., 37: 433. 22 Mai 1913. – Neotype (Samson & Gams, 1985): No. 104.13 (CBS).

Aspergillus tardus Bissett & Widden in Canad. J. Bot. 62: 2521. 19 Dec 1984. – Holotype: No. 183872 (DAOM).

Aspergillus tatenoi Y. Horie & al. in Trans. Mycol. Soc. Japan 33: 395. 1992. – Holotype: No. FA 0022 p.p. (CBM).

Aspergillus terreus Thom in Amer. J. Bot. 5: 85. Feb 1918. – Neotype (Samson & Gams, 1985): No. 17294 (IMI).

Aspergillus terricola E. J. Marchal in Rev. Mycol. (Toulouse) 15: 101. 1893. – Neotype (Samson & Gams, 1985): No. 172294 (IMI).

Aspergillus tetrazonus Samson & W. Gams in Samson & Pitt, Adv. Penicillium Aspergillus Syst.: 48. 1985. – Holotype: No. 89351 p.p. (IMI).

Aspergillus thermomutatus (Paden) S. W. Peterson in Mycol. Res. 96: 549. Jul 1992. – Basionym: *Aspergillus fischeri* var. *thermomutatus* Paden in Mycopathol. Mycol. Appl. 36: 161. 13 Nov 1968. – Holotype: No. 1108305 p.p. (BPI).

Aspergillus togoënsis (Henn.) Samson & Seifert in Samson & Pitt, Adv. Penicillium Aspergillus Syst.: 419. 1985. – Basionym: *Stilbothamnium togoënse* Henn. in Bot. Jahrb. Syst. 23: 542. 25 Mai 1897. – Neotype (Samson & Seifert, 1985): Zaire, *Louis 6190*, No. B 1009 (BR).

Aspergillus tonophilus Ohtsuki in Bot. Mag. (Tokyo) 75: 438. 25 Nov 1962. – Neotype (Samson & Gams, 1985): No. 108299 p.p. (IMI).

Aspergillus undulatus H. Z. Kong & Z. T. Qi in Acta Mycol. Sin. 4: 211. Nov 1985. – Holotype: No. 47644 p.p. (HMAS).

Aspergillus unguis (Emile-Weil & L. Gaudin) Thom & Raper in Mycologia 31: 667. Nov-Dec 1939. – Basionym: *Sterigmatocystis unguis* Emile-Weil & L. Gaudin in Arch. Méd. Exp. Anat. Pathol. 28: 463. 1918. – Neotype (Samson & Gams, 1985): No. 136526 (IMI).

Aspergillus unilateralis Thrower in Austral. J. Bot. 2: 355. Nov 1954. – Neotype (Samson & Gams, 1985): No. 62876 (IMI).

Aspergillus ustus (Bainier) Thom & Church, Aspergilli: 152. Mar 1926. – Basionym: *Sterigmatocystis usta* Bainier in Bull. Soc. Bot. France 28: 78. 1881. – Neotype (Samson & Gams, 1985): No. 211805 (IMI).

Aspergillus varians Wehmer in Bot. Centralbl. 80: 460. 13 Dec 1899. – Neotype (Samson & Gams, 1985): No. 172297 (IMI).

Aspergillus versicolor (Vuill.) Tirab. in Ann. Bot. (Roma) 7: 9. 31 Aug 1908. – Basionym: *Sterigmatocystis versicolor* Vuill. in Mirsky, Causes Err. Dét. Aspergill.: 15. 1903. – Neotype (Samson & Gams, 1985): No. 538.65 (CBS).

Aspergillus violaceobrunneus Samson & W. Gams in Samson & Pitt, Adv. Penicillium Aspergillus Syst.: 53. 1985. – Holotype: No. 61449 p.p. (IMI).

Aspergillus violaceofuscus Gasperini in Atti Soc. Tosc. Sci. Nat. Pisa, Processi Verbali 8: 326. 1887. – Neotype (Samson & Gams, 1985): No. 123.27 (CBS).

Aspergillus viridinutans Ducker & Thrower in Austral. J. Bot. 2: 355. Nov 1954. – Neotype (Samson & Gams, 1985): No. 62875 (IMI).

Aspergillus vitellinus (Massee) Samson & Seifert in Samson & Pitt, Adv. Penicillium Aspergillus Syst.: 417. 1985. – Basionym: *Sterigmatocystis vitellina* Ridl. ex Massee in J. Bot. 34: 152. Apr 1896. – Holotype: Singapore, 1894, *Ridley 2970* (K).

Aspergillus vitis Novobr. in Novosti Sist. Nizš. Rast. 9: 175. 4 Aug 1972. – Neotype (Kozakiewicz, 1989): No. 174724 (IMI).

Aspergillus warcupii Samson & W. Gams in Samson & Pitt, Adv. Penicillium Aspergillus Syst.: 50. 1985. – Holotype: No. 75885 p.p. (IMI).

Aspergillus wentii Wehmer in Centralbl. Bakteriol., 2. Abth., 2: 149. 27 Mar 1896. – Neotype (Samson & Gams, 1985): No. 17295 (IMI).

Aspergillus xerophilus Samson & Mouch. in Antonie van Leeuwenhoek J. Microbiol. Serol. 41: 348. 1975. – Holotype: No. 938.73 p.p. (CBS).

Aspergillus zonatus Kwon-Chung & Fennell in Raper & Fennell, Genus Aspergillus: 377. 1965. – Neotype (Samson & Gams, 1985): No. 506.65 (CBS).

Byssochlamys Westling
(holomorphs)

Byssochlamys fulva Olliver & G. Sm. in J. Bot. 71: 196. Jul 1933. – Neotype (designated here): No. 40021 (IMI). [Anamorph: *Paecilomyces fulvus* Stolk & Samson].

Byssochlamys nivea Westling in Svensk Bot. Tidskr. 3: 134. 28 Jun 1909. – Neotype (designated here): No. 100.11 (CBS). [Anamorph: *Paecilomyces niveus* Stolk & Samson].

Byssochlamys verrucosa Samson & Tansey in Trans. Brit. Mycol. Soc. 65: 512. Dec 1975. – Holotype: No. 605.74 (CBS). [Anamorph: *Paecilomyces verrucosus* Samson & Tansey].

Byssochlamys zollerniae C. Ram in Nova Hedwigia 16: 312. 5 Dec 1968. – Lectotype (designated here): icon in Nova Hedwigia 16: tab. 107. 5 Dec 1968. [Anamorph: *Paecilomyces zollerniae* Stolk & Samson].

Chaetosartorya Subram.
(holomorphs)

Chaetosartorya chrysella (Kwon-Chung & Fennell) Subram. in Curr. Sci. 41: 761. 5 Nov 1972. – Basionym: *Aspergillus chrysellus* Kwon-Chung & Fennell in Raper & Fennell, Genus Aspergillus: 424. 1965 (nom. holomorph.). – Neotype (Samson & Gams, 1985): No. 238612 (IMI). [Anamorph: *Aspergillus chryseides* Samson & W. Gams].

Chaetosartorya cremea (Kwon-Chung & Fennell) Subram. in Curr. Sci. 41: 761. 5 Nov 1972. – Basionym: *Aspergillus cremeus* Kwon-Chung & Fennell in Raper & Fennell, Genus Aspergillus: 418. 1965 (nom. holomorph.). – Neotype (Samson & Gams, 1985): No. 123749ii (IMI). [Anamorph: *Aspergillus cremeoflavus* Samson & W. Gams].

Chaetosartorya stromatoides B. J. Wiley & E. G. Simmons in Mycologia 65: 935. 25 Sep 1973. – Holotype: No. 8944 (QM). [Anamorph: *Aspergillus stromatoides* Raper & Fennell].

Cristaspora Fort & Guarro
(holomorphs)

Cristaspora arxii Fort & Guarro in Mycologia 76: 1115. 5 Dec 1984. – Holotype: No. 525.83 (CBS).

Dendrosphaera Pat.
(holomorphs)

Dendrosphaera eberhardtii Pat. in Bull. Soc. Mycol. France 23: 69. 1907. – Type not known.

Dichlaena Durieu & Mont.
(holomorphs)

Dichlaena lentisci Durieu & Mont. in Durieu, Expl. Sci. Algérie,

Bot., 1: 405. 1849. – Holotype: *von Höhnel* (FH).

Dichotomomyces D. B. Scott
(holomorphs)

Dichotomomyces cejpii (Milko) D. B. Scott in Trans. Brit. Mycol. Soc. 55: 313. 19 Oct 1970. – Basionym: *Talaromyces cejpii* Milko in Novosti Sist. Nizš. Rast. 1964: 208. 12 Aug 1964. – Neotype (designated here): No. 157.66 (CBS).

Emericella Berk.
(holomorphs)

Emericella acristata (Fennell & Raper) Y. Horie in Trans. Mycol. Soc. Japan 21: 491. Dec 1980. – Basionym: *Aspergillus nidulans* var. *acristatus* Fennell & Raper in Mycologia 47: 79. 8 Mar 1955 (nom. holomorph.). – Neotype (designated here): No. 61453 (IMI).

Emericella astellata (Fennell & Raper) Y. Horie in Trans. Mycol. Soc. Japan 21: 491. Dec 1980. – Basionym: *Aspergillus variecolor* var. *astellatus* Fennell & Raper in Mycologia 47: 81. 8 Mar 1955 (nom. holomorph.). – Neotype (designated here): No. 61455 (IMI).

Emericella aurantiobrunnea (G. A. Atkins & al.) Malloch & Cain in Canad. J. Bot. 50: 61. 8 Feb 1972. – Basionym: *Emericella nidulans* var. *aurantiobrunnea* G. A. Atkins & al. in Trans. Brit. Mycol. Soc. 41: 504. 19 Dec 1958. – Holotype: No. 74897 (IMI).

Emericella bicolor M. Chr. & States in Mycologia 70: 337. 25 Mai 1978. – Holotype: No. RMF 2058 (NY). [Anamorph: *Aspergillus bicolor* M. Chr. & States].

Emericella corrugata Udagawa & Y. Horie in Mycotaxon 4: 535. 18 Dec 1976. – Holotype: No. 2763 (NHL). [Anamorph: *Aspergillus corrugatus* Udagawa & Y. Horie].

Emericella dentata (D. K. Sandhu & R. S. Sandhu) Y. Horie in Trans. Mycol. Soc. Japan 21: 491. Dec 1980. – Basionym: *Aspergillus nidulans* var. *dentatus* D. K. Sandhu & R. S. Sandhu in Mycologia 55: 297. 7 Jun 1963 (nom. holomorph.). – Neotype (designated here): No. 126693 (IMI).

Emericella desertorum Samson & Mouch. in Antonie van Leeuwenhoek J. Microbiol. Serol. 40: 121. 1974. – Holotype: No. 653.73 (CBS).

Emericella echinulata (Fennell & Raper) Y. Horie in Trans. Mycol. Soc. Japan 21: 492. Dec 1980. – Basionym: *Aspergillus nidulans* var. *echinulatus* Fennell & Raper in Mycologia 47: 79. 8 Mar 1955 (nom. holomorph.). – Neotype (designated here): No. 61454 (IMI).

Emericella falconensis Y. Horie & al. in Trans. Mycol. Soc. Japan 30: 257. Oct 1989. – Holotype: No. 10001 p.p. (CBM). [Anamorph: *Aspergillus falconensis* Y. Horie].

Emericella foeniculicola Udagawa in Trans. Mycol. Soc. Japan 20: 13. Jun 1979. – Holotype: No. 2777 (NHL). [Anamorph: *Aspergillus foeniculicola* Udagawa].

Emericella foveolata Y. Horie in Trans. Mycol. Soc. Japan 19: 313. Oct 1978. – Holotype: No. 4547 (IFM). [Anamorph: *Aspergillus foveolatus* Y. Horie].

Emericella fruticulosa (Raper & Fennell) Malloch & Cain in Canad. J. Bot. 50: 61. 8 Feb 1972. – Basionym: *Aspergillus fructiculosus* Raper & Fennell, Genus Aspergillus: 506. 1965 (nom. holomorph.). – Neotype (Samson & Gams, 1985): No. 139279 (IMI). [Anamorph: *Aspergillus fruticans* Samson & W. Gams].

Emericella heterothallica (Kwon-Chung & al.) Malloch & Cain in Canad. J. Bot. 50: 62. 8 Feb 1972. – Basionym: *Aspergillus heterothallicus* Kwon-Chung & al. in Raper & Fennell, Genus Aspergillus: 502. 1965 (nom. holomorph.). – Neotype (Samson & Gams, 1985): No. 488.65 x 489.65 (CBS). [Anamorph: *Aspergillus compatibilis* Samson & W. Gams].

Emericella navahoënsis M. Chr. & States in Mycologia 74: 226. 2 Apr 1982. – Holotype: No. SD-5 (NY). [Anamorph: *Aspergillus navahoënsis* M. Chr. & States].

Emericella nidulans (Eidam) Vuill. in Compt. Rend. Hebd. Séances Acad. Sci. 184: 137. 1927. – Basionym: *Sterigmatocystis nidulans* Eidam in Beitr. Biol. Pflan-

zen 3: 393. 1883 (nom. holomorph.). – Neotype (Samson & Gams, 1985): No. 86806 (IMI). [Anamorph: *Aspergillus nidulans* (Eidam) G. Winter].

Emericella parvathecia (Raper & Fennell) Malloch & Cain in Canad. J. Bot. 50: 62. 8 Feb 1972. – Basionym: *Aspergillus parvathecius* Raper & Fennell, Genus Aspergillus: 509. 1965 (nom. holomorph.). – Neotype (Samson & Gams, 1985): No. 139280 (IMI). [Anamorph: *Aspergillus microthecius* Samson & W. Gams].

Emericella purpurea Samson & Mouch. in Antonie van Leeuwenhoek J. Microbiol. Serol. 41: 350. 1975. – Holotype: No. 754.74 (CBS). [Anamorph: *Aspergillus purpureus* Samson & Mouch.].

Emericella quadrilineata (Thom & Raper) C. R. Benj. in Mycologia 47: 680. 7 Oct 1955. – Basionym: *Aspergillus quadrilineatus* Thom & Raper in Mycologia 31: 660. Nov-Dec 1939 (nom. holomorph.). – Neotype (Samson & Gams, 1985): No. 89351 (IMI). [Anamorph: *Aspergillus tetrazonus* Samson & W. Gams].

Emericella rugulosa (Thom & Raper) C. R. Benj. in Mycologia 47: 680. 7 Oct 1955. – Basionym: *Aspergillus rugulosus* Thom & Raper in Mycologia 31: 660. Nov-Dec 1939 (nom. holomorph.). – Neotype (Samson & Gams, 1985): No. 136775 (IMI). [Anamorph: *Aspergillus rugulovalvus* Samson & W. Gams].

Emericella similis Y. Horie & al. in Trans. Mycol. Soc. Japan 31: 425. Dec 1990. – Holotype: No. 10007 (CBM).

Emericella spectabilis M. Chr. & Raper in Mycologia 70: 333. 25 Mai 1978. – Holotype: No. RMFH 429 (NY). [Anamorph: *Aspergillus spectabilis* M. Chr. & Raper].

Emericella striata (J. N. Rai & al.) Malloch & Cain in Canad. J. Bot. 50: 62. 8 Feb 1972. – Basionym: *Aspergillus striatus* J. N. Rai & al. in Canad. J. Bot. 42: 1521. 1964 (nom. holomorph.). – Neotype (Samson & Gams, 1985): No. 96679 (IMI). [Anamorph: *Aspergillus striatulus* Samson & W. Gams].

Emericella sublata Y. Horie in Trans. Mycol. Soc. Japan 20: 481. Dec 1979. – Holotype: No. 4553 (IFM). [Anamorph: *Aspergillus sublatus* Y. Horie].

Emericella undulata H. Z. Kong & Z. T. Qi in Acta Mycol. Sin. 4: 211. Nov 1985. – Holotype: No. 47644 (HMAS). [Anamorph: *Aspergillus undulatus* H. Z. Kong & Z. T. Qi].

Emericella unguis Malloch & Cain in Canad. J. Bot. 50: 62. 8 Feb 1972. – Holotype: No. 2393 (NRRL). [Anamorph: *Aspergillus unguis* (Emile-Weil & L. Gaudin) Thom & Raper].

Emericella variecolor Berk. & Broome in Berkeley, Intr. Crypt. Bot.: 340. Mar-Jun 1857. – Holotype: India, Sacundrabad, 3 Jul 1855, *s.coll.* (K). [Anamorph: *Aspergillus stellifer* Samson & W. Gams].

Emericella violacea (Fennell & Raper) Malloch & Cain in Canad. J. Bot. 50: 62. 8 Feb 1972. – Basionym: *Aspergillus violaceus* Fennell & Raper in Mycologia 47: 75. 8 Mar 1955 (nom. holomorph.). – Neotype (Samson & Gams, 1985): No. 61449 (IMI). [Anamorph: *Aspergillus violaceobrunneus* Samson & W. Gams].

Eupenicillium F. Ludw.
(holomorphs)

Eupenicillium abidjanum Stolk in Antonie van Leeuwenhoek J. Microbiol. Serol. 34: 49. 1968. – Holotype: No. 247.67 p.p. (CBS). [Anamorph: *Penicillium abidjanum* Stolk].

Eupenicillium alutaceum D. B. Scott in Mycopathol. Mycol. Appl. 36: 17. 31 Oct 1968. – Holotype: No. 317.67 p.p. (CBS). [Anamorph: *Penicillium alutaceum* D. B. Scott].

Eupenicillium anatolicum Stolk in Antonie van Leeuwenhoek J. Microbiol. Serol. 34: 46. 1968. – Holotype: No. 479.66 p.p. (CBS). [Anamorph: *Penicillium anatolicum* Stolk].

Eupenicillium angustiporcatum Takada & Udagawa in Trans. Mycol. Soc. Japan 24: 143. Jul 1983. – Holotype: No. 6481 p.p. (NHL). [Anamorph: *Penicillium angustiporcatum* Takada & Udagawa].

Eupenicillium baarnense (J. F. H. Beyma) Stolk & D. B. Scott in Persoonia 4: 401. 1 Aug 1967. – Basionym: *Penicillium baarnense* J. F. H. Beyma in Antonie van Leeuwenhoek J. Microbiol. Serol. 6: 271. 1940 (nom. holomorph.). – Neotype (Pitt, 1980): No. 134.41 (CBS). [Anamorph: *Penicillium vanbeymae* Pitt].

Eupenicillium brefeldianum (B. O. Dodge) Stolk & D. B. Scott in Persoonia 4: 400. 1 Aug 1967. – Basionym: *Penicillium brefeldianum* B. O. Dodge in Mycologia 25: 92. Mar-Apr 1933 (nom. holomorph.). – Neotype (Stolk & D. B. Scott, 1967): No. 216895 (IMI). [Anamorph: *Penicillium dodgei* Pitt].

Eupenicillium catenatum D. B. Scott in Mycopathol. Mycol. Appl. 36: 24. 31 Oct 1968. – Holotype: No. 352.67 p.p. (CBS). [Anamorph: *Penicillium catenatum* D. B. Scott].

Eupenicillium cinnamopurpureum D. B. Scott & Stolk in Antonie van Leeuwenhoek J. Microbiol. Serol. 33: 308. 1967. – Holotype (Pitt, 1980): No. 114483 (IMI). [Anamorph: *Penicillium phoeniceum* J. F. H. Beyma].

Eupenicillium crustaceum F. Ludw., Lehrb. Nied. Krypt.: 263. Jul 1892. – Lectotype (Stolk & D. B. Scott, 1967): icon of '*P. glaucum*' in Brefeld, Bot. Unters. Schimmelpilze 2: fig. 10-54. Feb 1874. [Anamorph: *Penicillium gladioli* L. McCulloch & Thom].

Eupenicillium cryptum Goch. in Mycotaxon 26: 349. 15 Jul 1986. – Holotype: No. 769 p.p. (NY). [Anamorph: *Penicillium cryptum* Goch.].

Eupenicillium egyptiacum (J. F. H. Beyma) Stolk & D. B. Scott in Persoonia 4: 401. 1 Aug 1967. – Basionym: *Penicillium egyptiacum* J. F. H. Beyma in Zentralbl. Bakteriol., 2. Abt., 88: 137. 24 Apr 1933 (nom. holomorph.). – Neotype (Pitt, 1980): No. 40580 p.p. (IMI). [Anamorph: *Penicillium nilense* Pitt].

Eupenicillium ehrlichii (Kleb.) Stolk & D. B. Scott in Persoonia 4: 400. 1 Aug 1967. – Basionym: *Penicillium ehrlichii* Kleb. in Ber. Deutsch. Bot. Ges. 48: 374. 29 Dec 1930 (nom. holomorph.). – Neotype (Pitt, 1980): No. 39737 (IMI). [Anamorph: *Penicillium klebahnii* Pitt].

Eupenicillium erubescens D. B. Scott in Mycopathol. Mycol. Appl. 36: 14. 31 Oct 1968. – Holotype: No. 318.67 p.p. (CBS). [Anamorph: *Penicillium erubescens* D. B. Scott].

Eupenicillium fractum Udagawa in Trans. Mycol. Soc. Japan 9: 51. 1 Nov 1968. – Holotype: No. 6104 p.p. (NHL). [Anamorph: *Penicillium fractum* Udagawa].

Eupenicillium gracilentum Udagawa & Y. Horie in Trans. Mycol. Soc. Japan 14: 373. 20 Dec 1973. – Holotype: No. 6452 p.p. (NHL). [Anamorph: *Penicillium gracilentum* Udagawa & Y. Horie].

Eupenicillium hirayamae D. B. Scott & Stolk in Antonie van Leeuwenhoek J. Microbiol. Serol. 33: 305. 1967. – Holotype: No. 229.60 (CBS). [Anamorph: *Penicillium hirayamae* Udagawa].

Eupenicillium inusitatum D. B. Scott in Mycopathol. Mycol. Appl. 36: 20. 31 Oct 1968. – Holotype: No. 351.67 p.p. (CBS). [Anamorph: *Penicillium inusitatum* D. B. Scott].

Eupenicillium javanicum (J. F. H. Beyma) Stolk & D. B. Scott in Persoonia 4: 398. 1 Aug 1967. – Basionym: *Penicillium javanicum* J. F. H. Beyma in Verh. Kon. Ned. Akad. Wetensch., Afd. Natuurk., Tweede Sect., 26(4): 17. 1929 (nom. holomorph.). – Neotype (Pitt, 1980): No. 39733 p.p. (IMI). [Anamorph: *Penicillium indonesiae* Pitt].

Eupenicillium katangense Stolk in Antonie van Leeuwenhoek J. Microbiol. Serol. 34: 42. 1968. – Holotype: No. 247.67 p.p. (CBS). [Anamorph: *Penicillium katangense* Stolk].

Eupenicillium lapidosum D. B. Scott & Stolk in Antonie van Leeuwenhoek J. Microbiol. Serol. 33: 298. 1967. – Holotype: No. 343.48 (CBS). [Anamorph: *Penicillium lapidosum* Raper & Fennell].

Eupenicillium lassenii Paden in Mycopathol. Mycol. Appl. 43: 266. 25 Mar 1971. – Holotype: No. JWP 69-26 p.p. (UVIC). [Anamorph: *Penicillium lassenii* Paden].

Eupenicillium levitum (Raper & Fennell) Stolk & D. B. Scott in Persoonia 4: 402. 1 Aug 1967. – Basionym: *Penicillium levitum* Raper & Fennell in Mycologia 40: 511. Sep-Oct 1948 (nom. holomorph.). – Neotype (Pitt, 1980): No. 39735 p.p. (IMI). [Anamorph: *Penicillium rasile* Pitt].

Eupenicillium limoneum Goch. & Zlattner in Stud. Mycol. 23: 100. 15 Mar 1983. – Holotype: No. 650.82 (CBS). [Anamorph: *Torulomyces lagena* Delitsch].

Eupenicillium lineolatum Udagawa & Y. Horie in Mycotaxon 5: 493. 6 Mai 1977. – Holotype: No. 2776 p.p. (NHL). [Anamorph: *Penicillium lineolatum* Udagawa & Y. Horie].

Eupenicillium ludwigii Udagawa in Trans. Mycol. Soc. Japan 10: 2. 1 Aug 1969. – Holotype: No. 6118 p.p. (NHL). [Anamorph: *Penicillium ludwigii* Udagawa].

Eupenicillium meliforme Udagawa & Y. Horie in Trans. Mycol. Soc. Japan 14: 376. 20 Dec 1973. – Holotype: No. 6468 p.p. (NHL). [Anamorph: *Penicillium meliforme* Udagawa & Y. Horie].

Eupenicillium meridianum D. B. Scott in Mycopathol. Mycol. Appl. 36: 12. 31 Oct 1968. – Holotype: No. 314.67 p.p. (CBS). [Anamorph: *Penicillium meridianum* D. B. Scott].

Eupenicillium molle Malloch & Cain in Canad. J. Bot. 50: 62. 8 Feb 1972. – Holotype: No. 45714

(TRTC). [Anamorph: *Penicillium molle* Pitt].

Eupenicillium nepalense Takada & Udagawa in Trans. Mycol. Soc. Japan 24: 146. Jul 1983. – Holotype: No. 6482 p.p. (NHL). [Anamorph: *Penicillium nepalense* Takada & Udagawa].

Eupenicillium ochrosalmoneum D. B. Scott & Stolk in Antonie van Leeuwenhoek J. Microbiol. Serol. 33: 302. 1967. – Holotype: No. 489.66 (CBS). [Anamorph: *Penicillium ochrosalmoneum* Udagawa].

Eupenicillium ornatum Udagawa in Trans. Mycol. Soc. Japan 9: 49. 1 Nov 1968. – Holotype: No. 6101 p.p. (NHL). [Anamorph: *Penicillium ornatum* Udagawa].

Eupenicillium osmophilum Stolk & Veenb.-Rijks in Antonie van Leeuwenhoek J. Microbiol. Serol. 40: 1. 1974. – Holotype: No. 462.72 p.p. (CBS). [Anamorph: *Penicillium osmophilum* Stolk & Veenb.-Rijks].

Eupenicillium parvum (Raper & Fennell) Stolk & D. B. Scott in Persoonia 4: 402. 1 Aug 1967. – Basionym: *Penicillium parvum* Raper & Fennell in Mycologia 40: 508. Sep-Oct 1948 (nom. holomorph.). – Neotype (Pitt, 1980): No. 359.48 (CBS). [Anamorph: *Penicillium papuanum* Udagawa & Y. Horie].

Eupenicillium pinetorum Stolk in Antonie van Leeuwenhoek J. Microbiol. Serol. 34: 37. 1968. – Holotype: No. 295.62 (CBS).

[Anamorph: *Penicillium pinetorum* M. Chr. & Backus].

Eupenicillium reticulisporum Udagawa in Trans. Mycol. Soc. Japan 9: 52. 1 Nov 1968. – Holotype: No. 6105 p.p. (NHL). [Anamorph: *Penicillium reticulisporum* Udagawa].

Eupenicillium rubidurum Udagawa & Y. Horie in Trans. Mycol. Soc. Japan 14: 381. 20 Dec 1973. – Holotype: No. 6460 p.p. (NHL). [Anamorph: *Penicillium rubidurum* Udagawa & Y. Horie].

Eupenicillium senticosum D. B. Scott in Mycopathol. Mycol. Appl. 36: 5. 31 Oct 1968. – Holotype: No. 316.67 p.p. (CBS). [Anamorph: *Penicillium senticosum* D. B. Scott].

Eupenicillium shearii Stolk & D. B. Scott in Persoonia 4: 396. 1 Aug 1967. – Holotype: No. 290.48 p.p. (CBS). [Anamorph: *Penicillium shearii* Stolk & D. B. Scott].

Eupenicillium sinaicum Udagawa & S. Ueda in Mycotaxon 14: 266. 22 Jan 1982. – Holotype: No. 2894 p.p. (NHL). [Anamorph: *Penicillium sinaicum* Udagawa & S. Ueda].

Eupenicillium stolkiae D. B. Scott in Mycopathol. Mycol. Appl. 36: 8. 31 Oct 1968. – Holotype: No. 315.67 p.p. (CBS). [Anamorph: *Penicillium stolkiae* D. B. Scott].

Eupenicillium terrenum D. B. Scott in Mycopathol. Mycol. Appl. 36: 1. 31 Oct 1968. – Holotype: No.

313.67 p.p. (CBS). [Anamorph: *Penicillium terrenum* D. B. Scott].

Eupenicillium tularense Paden in Mycopathol. Mycol. Appl. 43: 262. 25 Mar 1971. – Holotype: No. JWP 68-31 p.p. (UVIC). [Anamorph: *Penicillium tularense* Paden].

Eupenicillium zonatum Hodges & J. J. Perry in Mycologia 65: 697. 19 Jul 1973. – Holotype: No. FSL 525 p.p. (BPI). [Anamorph: *Penicillium zonatum* Hodges & J. J. Perry].

Eurotium Link : Fr
(holomorphs)

Eurotium amstelodami L. Mangin in Ann. Sci. Nat., Bot., ser. 9, 10: 360. 1909. – Neotype (Samson & Gams, 1985): No. 518.65 (CBS). [Anamorph: *Aspergillus vitis* Novobr.].

Eurotium appendiculatum Blaser in Sydowia 28: 38. Dec 1976. – Holotype: No. 8286 (ZT). [Anamorph: *Aspergillus appendiculatus* Blaser].

Eurotium athecium (Raper & Fennell) Arx, Gen. Fungi Sporul. Pure Cult., ed. 2: 91. 1974. – Basionym: *Aspergillus athecius* Raper & Fennell, Genus Aspergillus: 183. 1965 (nom. holomorph.). – Neotype (Samson & Gams, 1985): No. 32048 (IMI). [Anamorph: *Aspergillus atheciellus* Samson & W. Gams].

Eurotium carnoyi Malloch & Cain in Canad. J. Bot. 50: 63. 8 Feb 1972. – Neotype (Samson & Gams, 1985): No. 172279 (IMI). [Anamorph: *Aspergillus neocarnoyi* Kozak.].

Eurotium chevalieri L. Mangin in Ann. Sci. Nat., Bot., ser. 9, 10: 361. 1909. – Neotype (Blaser, 1975): No. 211382 (IMI). [Anamorph: *Aspergillus chevalieri* L. Mangin].

Eurotium cristatum (Raper & Fennell) Malloch & Cain in Canad. J. Bot. 50: 64. 8 Feb 1972. – Basionym: *Aspergillus cristatus* Raper & Fennell, Genus Aspergillus: 169. 1965 (nom. holomorph.). – Neotype (Samson & Gams, 1985): No. 172280 (IMI). [Anamorph: *Aspergillus cristatellus* Kozak.].

Eurotium echinulatum Delacr. in Bull. Soc. Mycol. France 9: 266. 1893. – Neotype (Blaser, 1975): No. 211378 (IMI). [Anamorph: *Aspergillus brunneus* Delacr.].

Eurotium glabrum Blaser in Sydowia 28: 35. Dec 1976. – Holotype: No. 8218 T (ZT). [Anamorph: *Aspergillus glaber* Blaser].

Eurotium halophilicum C. M. Chr. & al. in Mycologia 51: 636. 17 Mar 1961. – Neotype (Samson & Gams, 1985): No. NRRL 2739 (BPI). [Anamorph: *Aspergillus halophilicus* C. M. Chr. & al.].

Eurotium herbariorum Link in Ges. Naturf. Freunde Berlin Mag. Neuesten Entdeck. Gesammten Naturk. 3: 31. 1809 : Fr. – Neotype (Malloch & Cain, 1972): No. 137960 (DAOM). [Anamorph: *Aspergillus glaucus* Link].

Eurotium intermedium Blaser in Sydowia 28: 41. Dec 1976. – Basionym: *Aspergillus chevalieri* var. *intermedius* Thom & Raper in Misc. Publ. U.S. Dept. Agric. 426: 21. 1941 (nom. holomorph.). – Neotype (Kozakiewicz, 1989): No. 89278 (IMI). [Anamorph: *Aspergillus intermedius* Blaser].

Eurotium leucocarpum Hadlok & Stolk in Antonie van Leeuwenhoek J. Microbiol. Serol. 35: 9. 1969. – Holotype: No. 353.68 (CBS). [Anamorph: *Aspergillus leucocarpus* Hadlok & Stolk].

Eurotium medium R. Meissn. in Bot. Zeitung, 2. Abt., 55: 356. 1 Dec 1897. – Neotype (Samson & Gams, 1985): No. 113.27 (CBS). [Anamorph: *Aspergillus medius* R. Meissn.].

Eurotium niveoglaucum (Thom & Raper) Malloch & Cain in Canad. J. Bot. 50: 64. 8 Feb 1972. – Basionym: *Aspergillus niveoglaucus* Thom & Raper in Misc. Publ. U.S. Dept. Agric. 426: 35. 1941 (nom. holomorph.). – Neotype (Samson & Gams, 1985): No. 32050ii (IMI). [Anamorph: *Aspergillus glauconiveus* Samson & W. Gams].

Eurotium pseudoglaucum (Blochwitz) Malloch & Cain in Canad. J. Bot. 50: 64. 8 Feb 1972. – Basionym: *Aspergillus pseudoglaucus* Blochwitz in Ann. Mycol. 27: 207. 25 Jun 1929 (nom. holomorph.). – Neotype (Samson & Gams, 1985): No. 16122ii (IMI). [Anamorph: *Aspergillus glaucoaffinis* Samson & W. Gams].

Eurotium repens de Bary in Abh. Senckenberg. Naturf. Ges. 7: 379. 1870. – Neotype (Samson & Gams, 1985): No. 529.65 (CBS). [Anamorph: *Aspergillus reptans* Samson & W. Gams].

Eurotium rubrum Jos. König & al. in Z. Untersuch. Nahrungs- Genussmittel 4: 726. 1901. – Neotype (Samson & Gams, 1985): No. 530.65 (CBS). [Anamorph: *Aspergillus rubrobrunneus* Samson & W. Gams].

Eurotium tonophilum Ohtsuki in Bot. Mag. (Tokyo) 75: 438. 25 Nov 1962. – Neotype (Samson & Gams, 1985): No. 108299 (IMI). [Anamorph: *Aspergillus tonophilus* Ohtsuki].

Eurotium xerophilum Samson & Mouch. in Antonie van Leeuwenhoek J. Microbiol. Serol. 41: 348. 1975. – Holotype: No. 938.73 (CBS). [Anamorph: *Aspergillus xerophilus* Samson & Mouch.].

Fennellia B. J. Wiley &
E. G. Simmons (holomorphs)

Fennellia flavipes B. J. Wiley & E. G. Simmons in Mycologia 65: 937. 25 Sep 1973. – Holotype: No. 9131 (QM). [Anamorph: *Aspergillus flavipes* (Bainier & Sartory) Thom & Church].

Fennellia monodii Locq.-Lin. in Mycotaxon 39: 10. 26 Nov 1990. – Holotype: No. 89-3570 LCP (PC).

Fennellia nivea (B. J. Wiley & E. G. Simmons) Samson in Stud. Mycol. 18: 5. 3 Jan 1979. – Basionym: *Emericella nivea* B. J. Wiley & E. G. Simmons in Mycologia 65: 934. 25 Sep 1973. – Holotype: No. 8942 (QM). [Anamorph: *Aspergillus niveus* Blochwitz].

Geosmithia Pitt
(anamorphs)

Geosmithia argillacea (Stolk & al.) Pitt in Canad. J. Bot. 57: 2026. 1 Oct 1979. – Basionym: *Penicillium argillaceum* Stolk & al. in Trans. Brit. Mycol. Soc. 53: 307. 8 Oct 1969. – Holotype: No. 101.69 (CBS).

Geosmithia cylindrospora (G. Sm.) Pitt in Canad. J. Bot. 57: 2024. 1 Oct 1979. – Basionym: *Penicillium cylindrosporum* G. Sm. in Trans. Brit. Mycol. Soc. 40: 483. 20 Dec 1957. – Neotype (Pitt, 1979): No. 71623 (IMI).

Geosmithia emersonii (Stolk) Pitt in Canad. J. Bot. 57: 2027. 1 Oct 1979. – Basionym: *Penicillium emersonii* Stolk in Antonie van Leeuwenhoek J. Microbiol. Serol. 31: 262. 1965. – Holotype (Stolk, 1965): No. 393.64 p.p. (CBS).

Geosmithia lavendula (Raper & Fennell) Pitt in Canad. J. Bot. 57: 2022. 1 Oct 1979. – Basionym: *Penicillium lavendulum* Raper & Fennell in Mycologia 40: 530. Sep-Oct 1948. – Neotype (Pitt, 1979): No. 40570 (IMI).

Geosmithia namyslowskii (K. M. Zalessky) Pitt in Canad. J. Bot. 57: 2024. 1 Oct 1979. – Basionym: *Penicillium namyslowskii* K. M. Zalessky in Bull. Int. Acad. Polon.

Sci., Cl. Sci. Math., Sér. B, Sci. Nat., 1927: 479. 1927. – Neotype (Pitt, 1979): No. 40033 (IMI).

Geosmithia putterillii (Thom) Pitt in Canad. J. Bot. 57: 2022. 1 Oct 1979. – Basionym: *Penicillium putterillii* Thom, Penicillia: 368. Jan 1930. – Neotype (Pitt, 1979): No. 40212 (IMI).

Geosmithia swiftii Pitt in Canad. J. Bot. 57: 2028. 1 Oct 1979. – Holotype: No. 40045 (IMI).

Geosmithia viridis Pitt & A. D. Hocking in Mycologia 77: 822. 15 Oct 1985. – Holotype: No. 1863 (FRR).

Hemicarpenteles A. K. Sarbhoy & Elphick (holomorphs)

Hemicarpenteles acanthosporus Udagawa & Takada in Bull. Natl. Sci. Mus. Tokyo, ser. 2, 14: 503. 30 Sep 1971. – Holotype: No. 22462 (NHL). [Anamorph: *Aspergillus acanthosporus* Udagawa & Takada].

Hemicarpenteles paradoxus A. K. Sarbhoy & Elphick in Trans. Brit. Mycol. Soc. 51: 156. 30 Mar 1968. – Neotype (Samson & Gams, 1985): No. 61446 (IMI). [Anamorph: *Aspergillus paradoxus* Fennell & Raper].

Merimbla Pitt
(anamorphs)

Merimbla ingelheimensis (J. F. H. Beyma) Pitt in Canad. J. Bot. 57: 2395. 1 Nov 1979. – Basionym: *Penicillium ingelheimense* J. F. H.

Beyma in Antonie van Leeuwenhoek J. Microbiol. Serol. 8: 109. 1942. – Neotype (Pitt & Hocking, 1979): No. 234977 (IMI).

Neosartorya Malloch & Cain (holomorphs)

Neosartorya aurata (Warcup) Malloch & Cain in Canad. J. Bot. 50: 2620. 26 Jan 1973. – Basionym: *Aspergillus auratus* Warcup in Raper & Fennell, Genus Aspergillus: 263. 1965 (nom. holomorph.). – Neotype (Kozakiewicz, 1989): No. 75886 (IMI). [Anamorph: *Aspergillus igneus* Kozak.].

Neosartorya aureola (Fennell & Raper) Malloch & Cain in Canad. J. Bot. 50: 2620. 26 Jan 1973. – Basionym: *Aspergillus aureolus* Fennell & Raper in Mycologia 47: 71. 8 Mar 1955 (nom. holomorph.). – Neotype (Samson & Gams, 1985): No. 105.55 (CBS). [Anamorph: *Aspergillus aureoluteus* Samson & W. Gams].

Neosartorya fennelliae Kwon-Chung & S. J. Kim in Mycologia 66: 629. 28 Aug 1974. – Neotype (Kozakiewicz, 1989): No. 278382 (IMI). [Anamorph: *Aspergillus fennelliae* Kwon-Chung & S. J. Kim].

Neosartorya fischeri (Wehmer) Malloch & Cain in Canad. J. Bot. 50: 2621. 26 Jan 1973. – Basionym: *Aspergillus fischeri* Wehmer in Centralbl. Bakteriol., 2. Abth., 18: 390. 22 Apr 1907 (nom. holomorph.). – Neotype (Samson & Gams, 1985): No. 21139ii (IMI). [Anamorph: *Aspergillus fischerianus* Samson & W. Gams].

Neosartorya glabra (Fennell & Raper) Kozak. in Mycol. Pap. 161: 56. 30 Jun 1989. – Basionym: *Aspergillus fischeri* var. *glaber* Fennell & Raper in Mycologia 47: 74. 8 Mar 1955 (nom. holomorph.). – Neotype (Samson & Gams, 1985): No. 61447 (IMI). [Anamorph: *Aspergillus neoglaber* (Fennell & Raper) Kozak.].

Neosartorya hiratsukae Udagawa & al. in Trans. Mycol. Soc. Japan 32: 23. Apr 1991. – Holotype: No. 3008 (NHL). [Anamorph: *Aspergillus hiratsukae* Udagawa & al.].

Neosartorya pseudofischeri S. W. Peterson in Mycol. Res. 96: 549. Jul 1992. – Holotype: 1108305 p.p. (BPI). [Anamorph: *Aspergillus thermomutatus* (Paden) S. W. Peterson].

Neosartorya quadricincta (E. Yuill) Malloch & Cain in Canad. J. Bot. 50: 2621. 26 Jan 1973. – Basionym: *Aspergillus quadricinctus* E. Yuill in Trans. Brit. Mycol. Soc. 36: 58. 28 Mar 1953 (nom. holomorph.). – Holotype: No. 48583ii (IMI). [Anamorph: *Aspergillus quadricingens* Kozak.].

Neosartorya spathulata Takada & Udagawa in Mycotaxon 24: 396. 11 Dec 1985. – Holotype: No. 2947 (NHL). [Anamorph: *Aspergillus spathulatus* Takada & Udagawa].

Neosartorya spinosa (Raper & Fennell) Kozak. in Mycol. Pap. 161: 58. 30 Jun 1989. – Basionym: *Aspergillus fischeri* var. *spinosus* Raper & Fennell, Genus Aspergillus: 256. 1965 (nom. holomorph.). – Neotype (Samson & Gams, 1985): No. 211390 (IMI). [Anamorph: *Aspergillus spinosus* Kozak.].

Neosartorya stramenia (R. Novak & Raper) Malloch & Cain in Canad. J. Bot. 50: 2622. 26 Jan 1973. – Basionym: *Aspergillus stramenius* R. Novak & Raper in Raper & Fennell, Genus Aspergillus: 260. 1965 (nom. holomorph.). – Neotype (Samson & Gams, 1985): No. 172293 (IMI). [Anamorph: *Aspergillus paleaceus* Samson & W. Gams].

Neosartorya tatenoi Y. Horie & al. in Trans. Mycol. Soc. Japan 33: 395. 1992. – Holotype: No. FA 0022 p.p. (CBM). [Anamorph: *Aspergillus tatenoi* Horie & al.].

Paecilomyces Bainier
(anamorphs)

Paecilomyces aegyptiacus S. Ueda & Udagawa in Trans. Mycol. Soc. Japan 24: 135. Jul 1983. – Holotype: No. 2914 (NHL).

Paecilomyces aerugineus Samson in Stud. Mycol. 6: 20. 10 Jun 1974. – Holotype: No. 350.66 (CBS).

Paecilomyces amoeneroseus (Henn.) Samson in Stud. Mycol. 6: 37. 10 Jun 1974. – Basionym: *Isaria amoenerosea* Henn. in Hedwigia 41: (66). 24 Apr 1902. – Holotype: Brazil, Tijuca, Apr 1897, *Ule* in herb. Hennings (B).

Paecilomyces breviramosus Bissett in S. Hughes, Fungi Canad.: n. 159. 1979. – Holotype: No. 154555 (DAOM).

Paecilomyces byssochlamydoides Stolk & Samson in Stud. Mycol. 2: 45. 1 Nov 1972. – Holotype: No. 413.71 p.p. (CBS).

Paecilomyces carneus (Duché & R. Heim) A. H. S. Br. & G. Sm. in Trans. Brit. Mycol. Soc. 40: 70. 25 Mar 1957. – Basionym: *Spicaria carnea* Duché & R. Heim in Trav. Cryptog. Louis L. Mangin: 454. Sep 1931. – Neotype (Samson, 1974): No. 239.32 (CBS).

Paecilomyces cateniannulatus Z. Q. Liang in Acta Phytopathol. Sin. 11: 10. Dec 1981. – Holotype: No. CGAC 2601 (HGAS).

Paecilomyces cateniobliquus Z. Q. Liang in Acta Phytopathol. Sin. 11: 9. Dec 1981. – Holotype: No. CGAC 2401 (HGAS).

Paecilomyces cicadae (Miq.) Samson in Stud. Mycol. 6: 52. 10 Jun 1974. – Basionym: *Isaria cicadae* Miq. in Bull. Sci. Phys. Nat. Néerl. 1838: 86. 15 Jun 1838. – Neotype (designated here): Malaysia, Rnggano, Boeha-boeha, *Lutjeharms*, No. 961245-154 (L).

Paecilomyces cinnamomeus (Petch) Samson & W. Gams in Stud. Mycol. 6: 62. 10 Jun 1974. – Basionym: *Verticillium cinnamomeum* Petch in Trans. Brit. Mycol. Soc. 16: 233. 2 Mai 1932. – Holotype: Missouri, Ocean Springs, 1920, *Miles* (BPI).

Paecilomyces clavisporus Hammill in Mycologia 62: 109. 31 Mar 1970. – Neotype (designated here): No. 329.70 (CBS).

Paecilomyces coleopterorum Samson & H. C. Evans in Stud. Mycol. 6: 47. 10 Jun 1974. – Holotype: No. R.S. 0172 (CBS).

Paecilomyces crustaceus Apinis & Chesters in Trans. Brit. Mycol. Soc. 47: 429. 23 Sep 1964. – Neotype (designated here): No. 102470 (IMI).

Paecilomyces farinosus (Holmsk.) A. H. S. Br. & G. Sm. in Trans. Brit. Mycol. Soc. 40: 50. 25 Mar 1957. – Basionym: *Ramaria farinosa* Holmsk. in Nye Saml. Kongel. Danske Vidensk. Selsk. Skr. 1: 279. 1781. – Neotype (designated here): ad Driesen Lasch, *Klotzsch*, No. 910.225-729 (L).

Paecilomyces fulvus Stolk & Samson in Persoonia 6: 354. 27 Dec 1971. – Holotype: No. 132.33 (CBS).

Paecilomyces fumosoroseus (Wize) A. H. S. Br. & G. Sm. in Trans. Brit. Mycol. Soc. 40: 67. 25 Mar 1957. – Basionym: *Isaria fumosorosea* Wize in Bull. Int. Acad. Sci. Cracovie, Cl. Sci. Math., 1904: 72. 1904. – Neotype (designated here): on dead glow worm, Wheatfen Broad, *Petch* (K).

Paecilomyces ghanensis Samson & H. C. Evans in Stud. Mycol. 6: 46. 10 Jun 1974. – Holotype: No. R.S. 0096 (CBS).

Paecilomyces gunnii Z. Q. Liang in Acta Mycol. Sin. 4: 163. Aug 1985. – Holotype: No. CGAC 29031 (HGAS).

Paecilomyces hawkesii Z. M. Xiao & al. in Acta Mycol. Sin. 3: 109. Mai 1984. – Holotype: No. 43719 (HMAS).

Paecilomyces inflatus (Burnside) J. W. Carmich. in Canad. J. Bot. 40: 1148. 27 Aug 1962. – Basionym: *Myceliophthora inflata* Burnside in Pap. Michigan Acad. Sci. 8: 82-84. 7 Apr 1927. – Neotype (Samson, 1974): No. 259.39 (CBS).

Paecilomyces javanicus (Frieder. & W. Bally) A. H. S. Br. & G. Sm. in Trans. Brit. Mycol. Soc. 40: 65. 25 Mar 1957. – Basionym: *Spicaria javanica* Frieder. & W. Bally in Koffiebessenboeboek-Fonds Meded. 6: 146. 1923. – Neotype (Samson, 1974): No. 134.22 (CBS).

Paecilomyces leycettanus (Evans & Stolk) Stolk & al. in Persoonia 6: 342. 27 Dec 1971. – Basionym: *Penicillium leycettanum* H. C. Evans & Stolk in Trans. Brit. Mycol. Soc. 56: 45. 29 Feb 1971. – Holotype: No. 398.68 (CBS).

Paecilomyces lilacinus (Thom) Samson in Stud. Mycol. 6: 58. 10 Jun 1974. – Basionym: *Penicillium lilacinum* Thom in U.S.D.A. Bur. Anim. Industr. Bull. 118: 73. 1910. – Neotype (designated here): No. 27830 (IMI).

Paecilomyces marquandii (Massee) S. Hughes in Mycol. Pap. 45:

30. 12 Dec 1951. – Basionym: *Verticillium marquandii* Massee in Trans. Brit. Mycol. Soc. 1: 24. 1898. – Lectotype (Hughes, 1951): Guernsey, 10 Sep 1897, *Marquand* (K).

Paecilomyces niphetodes Samson in Stud. Mycol. 6: 65. 10 Jun 1974. – Holotype: No. 229.73 (CBS).

Paecilomyces niveus Stolk & Samson in Persoonia 6: 351. 27 Dec 1971. – Holotype: No. 100.11 (CBS).

Paecilomyces nostocoides M. T. Dunn in Mycologia 75: 179. 24 Feb 1983. – Holotype: No. MTD 3153 (BPI).

Paecilomyces pascua Pitt & A. D. Hocking in Mycologia 77: 822. 15 Oct 1985. – Holotype: No. 1925 (FRR).

Paecilomyces penicillatus (Höhn.) Samson in Stud. Mycol. 6: 72. 10 Jun 1974. – Basionym: *Spicaria penicillata* Höhn. in Ann. Mycol. 2: 56. 15 Feb 1904. – Holotype: Wienerwald, 1903, *von Höhnel* (FH).

Paecilomyces puntonii (Vuill.) Nann., Repert. Mic. Uomo: 245. Jul-Dec 1934. – Basionym: *Corethropsis puntonii* Vuill. in Compt. Rend. Hebd. Séances Acad. Sci. 190: 1334. 1930. – Neotype (designated here): No. 250.37 (CBS).

Paecilomyces ramosus Samson & H. C. Evans in Stud. Mycol. 6: 44. 10 Jun 1974. – Holotype: No. R.S. 0177 (CBS).

Paecilomyces reniformis Samson & H. C. Evans in Stud. Mycol. 6: 43. 10 Jun 1974. – Holotype: No. R.S. 0029 (CBS).

Paecilomyces sinensis Q. T. Chen & al. in Acta Mycol. Sin. 3: 25. Feb 1984. – Holotype: No. 43720 (HMAS).

Paecilomyces suffultus (Petch) Samson in Stud. Mycol. 6: 55. 10 Jun 1974. – Basionym: *Cylindrodendrum suffultum* Petch in Trans. Brit. Mycol. Soc. 27: 91. 18 Apr 1944. – Holotype: Norwich, Trowse Marshes, *Ellis* (K).

Paecilomyces sulphurellus (Sacc.) Samson & W. Gams in Stud. Mycol. 6: 67. 10 Jun 1974. – Basionym: *Verticillium sulphurellum* Sacc., Fung. Ital.: tab. 641. Jul 1881. – Neotype (designated here): No. W.G. 1619 (CBS).

Paecilomyces tenuipes (Peck) Samson in Stud. Mycol. 6: 49. 10 Jun 1974. – Basionym: *Isaria tenuipes* Peck in Annual Rep. New York State Mus. 33: 49. Jun 1883. – Holotype: U.S.A., Center, *Peck* (NYS).

Paecilomyces variotii Bainier in Bull. Soc. Mycol. France 23: 26. 1907. – Neotype (designated here): No. 102.74 (CBS).

Paecilomyces verrucosus Samson & Tansey in Trans. Brit. Mycol. Soc. 65: 512. Dec 1975. – Holotype: No. 605.74 (CBS).

Paecilomyces viridis Segretain ex Samson in Stud. Mycol. 6: 64. 10 Jun 1974. – Holotype: No. 348.65 (CBS).

Paecilomyces xylariiformis (Lloyd) Samson in Stud. Mycol. 6: 54. 10 Jun 1974. – Basionym: *Isaria xylariiformis* Lloyd, Mycol. Not. 7: 1200. Jul 1923. – Holotype: Brazil, *Rick* in herb. Lloyd No. 42613 (BPI).

Paecilomyces zollerniae Stolk & Samson in Persoonia 6: 356. 27 Dec 1971. – Neotype (designated here): No. 374.70 (CBS).

Penicilliopsis Solms
(holomorphs)

Penicilliopsis africana Samson & Seifert in Samson & Pitt, Adv. Penicillium Aspergillus Syst.: 408. 1985. – Holotype: Metiquette, *Louis 6275* (BR). [Anamorph: *Stilbodendron cervinum* (Cooke & Massee) Samson & Seifert].

Penicilliopsis clavariiformis Solms in Ann. Jard. Bot. Buitenzorg 6: 53. 1887. – Lectotype (designated here; see also Samson & Seifert, 1985): Bot. Garden Bogor, *Solms-Laubach* in herb. Hauman (BR). [Anamorph: *Sarophorum palmicola* (Henn.) Seifert & Samson].

Penicillium Link : Fr.
(anamorphs)

Penicillium abidjanum Stolk in Antonie van Leeuwenhoek J. Microbiol. Serol. 34: 49. 1968. – Holotype: No. 247.67 p.p. (CBS).

Penicillium aculeatum Raper & Fennell in Mycologia 40: 535. Sep-Oct 1948. – Neotype (Pitt, 1980): No. 40588 (IMI).

Penicillium adametzii K. M. Zalessky in Bull. Int. Acad. Polon. Sci., Cl. Sci. Math., Sér. B, Sci. Nat., 1927: 507. 1927. – Neotype (Pitt, 1980): No. 39751 (IMI).

Penicillium aethiopicum Frisvad in Mycologia 81: 848. 11 Jan 1990. – Holotype: No. 285524 (IMI).

Penicillium allahabadense B. S. Mehrotra & D. Kumar in Canad. J. Bot. 40: 1399. 19 Oct 1962. – Neotype (designated here): No. 304.63 (CBS).

Penicillium allii Vincent & Pitt in Mycologia 81: 300. 19 Apr 1989. – Holotype: Vincent 114 (MU).

Penicillium alutaceum D. B. Scott in Mycopathol. Mycol. Appl. 36: 17. 31 Oct 1968. – Holotype: No. 317.67 p.p. (CBS).

Penicillium anatolicum Stolk in Antonie van Leeuwenhoek J. Microbiol. Serol. 34: 46. 1968. – Holotype: No. 479.66 p.p. (CBS).

Penicillium angustiporcatum Takada & Udagawa in Trans. Mycol. Soc. Japan 24: 143. Jul 1983. – Holotype: No. 6481 p.p. (NHL).

Penicillium ardesiacum Novobr. in Novosti Sist. Nizš. Rast. 11: 228. 11 Nov 1974. – Neotype (designated here): No. 174719 (IMI).

Penicillium arenicola Chalab. in Bot. Mater. Otd. Sporov. Rast. 6: 162. 28 Jun 1950. – Neotype (Pitt, 1980): No. 117658 (IMI).

Penicillium asperosporum G. Sm. in Trans. Brit. Mycol. Soc. 48: 275. 22 Jun 1965. (≡ *Penicillium echinosporum* G. Sm. in Trans. Brit. Mycol. Soc. 45: 387. 1962 [non Nehira in J. Ferment. Technol. 11: 849. 1933]). – Holotype: No. 80450 (IMI).

Penicillium assiutense Samson & Abdel-Fattah in Persoonia 9: 501. 13 Jul 1978. – Holotype: No. 147.78 p.p. (CBS).

Penicillium atramentosum Thom in U.S.D.A. Bur. Anim. Industr. Bull. 118: 65. 1910. – Neotype (Pitt, 1980): No. 39752 (IMI).

Penicillium atrovenetum G. Sm. in Trans. Brit. Mycol. Soc. 39: 112. 20 Mar 1956. – Neotype (designated here): No. 61837 (IMI).

Penicillium aurantiogriseum Dierckx in Ann. Soc. Sci. Bruxelles 25: 88. 1901. – Neotype (Pitt, 1980): No. 195050 (IMI).

Penicillium bilaiae Chalab. in Bot. Mater. Otd. Sporov. Rast. 6: 165. 28 Jun 1950. – Neotype (Pitt, 1980): No. 113677 (IMI).

Penicillium brasilianum Bat. in Anais Soc. Biol. Pernambuco 15: 162. 1957. – Holotype: No. IMUR 56 (URM).

Penicillium brevicompactum Dierckx in Ann. Soc. Sci. Bruxelles 25: 88. 1901. – Neotype (Pitt, 1980): No. 40225 (IMI).

Penicillium brunneum Udagawa in J. Agric. Sci. Tokyo Nogyo Daigaku 5: 16. 1959. – Holotype: No. 6054 (NHL).

Penicillium camemberti Thom in U.S.D.A. Bur. Anim. Industr. Bull. 82: 33. 1906. – Neotype (Pitt, 1980): No. 27831 (IMI).

Penicillium canescens Sopp in Skr. Vidensk.-Selsk. Christiana, Math.-Naturvidensk. Kl. 11: 181. 1912. – Neotype (Pitt, 1980): No. 28260 (IMI).

Penicillium capsulatum Raper & Fennell in Mycologia 40: 528. Sep-Oct 1948. – Neotype (Pitt, 1980): No. 40576 (IMI).

Penicillium catenatum D. B. Scott in Mycopathol. Mycol. Appl. 36: 24. 31 Oct 1968. – Holotype: No. 352.67 p.p. (CBS).

Penicillium chalybeum Pitt & A. D. Hocking in Mycotaxon 22: 204. 8 Feb 1985. – Holotype: No. 2660 (FRR).

Penicillium chermesinum Biourge in Cellule 33: 284. 1923. – Neotype (Pitt, 1980): No. 191730 (IMI).

Penicillium chrysogenum Thom in U.S.D.A. Bur. Anim. Industr. Bull. 118: 58. 1910. – Neotype (Pitt, 1980): No. 24314 (IMI).

Penicillium cinerascens Biourge in Cellule 33: 308. 1923. – Neotype (designated here): No. 92234 (IMI).

Penicillium citreonigrum Dierckx in Ann. Soc. Sci. Bruxelles 25: 86. 1901. – Neotype (Pitt, 1980): No. 92209i (IMI).

Penicillium citrinum Thom in U.S.D.A. Bur. Anim. Industr. Bull. 118: 61. 1910. – Neotype (Pitt, 1980): No. 92196ii (IMI).

Penicillium clavigerum Demelius in Verh. Zool.-Bot. Ges. Wien 72: 74. 1923. – Neotype (designated here): No. 39807 (IMI).

Penicillium coalescens Quintan. in Mycopathologia 84: 115. 1983. – Neotype (designated here): No. 103.83 (CBS).

Penicillium commune Thom in U.S.D.A. Bur. Anim. Industr. Bull. 118: 56. 1910. – Neotype (designated here): No. 39812 (IMI).

Penicillium concentricum Samson & al. in Stud. Mycol. 11: 17. 30 Jan 1976. – Holotype: No. 477.75 (CBS).

Penicillium confertum (Frisvad & al.) Frisvad in Mycologia 81: 852. 11 Jan 1990. – Basionym: *Penicillium glandicola* var. *confertum* Frisvad & al. in Canad. J. Bot. 65: 769. 23 Apr 1987. – Holotype (Frisvad & al., 1987): No. 296930 (IMI).

Penicillium coprobium Frisvad in Mycologia 81: 853. 11 Jan 1990. – Holotype: No. 293209 (IMI).

Penicillium coprophilum (Berk. & M. A. Curtis) Seifert & Samson in Samson & Pitt, Adv. Penicillium Aspergillus Syst.: 145. 1985. – Basionym: *Coremium coprophilum* Berk. & M. A. Curtis in J. Linn. Soc., Bot., 10: 363. 16 Jun 1868. – Holotype: Cuba, *Wright 666* (K).

Penicillium coralligerum Nicot & Pionnat in Bull. Soc. Mycol. France 78: 245. 10 Jan 1963 ["1962"]. – Neotype (designated here): No. 99159 (IMI).

Penicillium corylophilum Dierckx in Ann. Soc. Sci. Bruxelles 25: 86. 1901. – Neotype (Pitt, 1980): No. 39754 (IMI).

Penicillium crateriforme J. C. Gilman & E. V. Abbott in Iowa State Coll. J. Sci. 1: 293. 1 Apr 1927. – Neotype (designated here): No. 94165 (IMI).

Penicillium cremeogriseum Chalab. in Bot. Mater. Otd. Sporov. Rast. 6: 168. 28 Jun 1950. – Neotype (designated here): No. 223.66 (CBS).

Penicillium crustosum Thom, Penicillia: 399. Jan 1930. – Neotype (Pitt, 1980): No. 91917 (IMI).

Penicillium cryptum Goch. in Mycotaxon 26: 349. 15 Jul 1986. – Holotype: No. 769 p.p. (NY).

Penicillium cyaneum (Bainier & Sartory) Biourge in Cellule 33: 102. 1923. – Basionym: *Citromyces cyaneus* Bainier & Sartory in Bull. Soc. Mycol. France 29: 157. 1913. – Neotype (Pitt, 1980): No. 39744 (IMI).

Penicillium daleae K. M. Zalessky in Bull. Int. Acad. Polon. Sci., Cl. Sci. Math., Sér. B, Sci. Nat., 1927: 495. 1927. – Neotype (Pitt, 1980): No. 89338 (IMI).

Penicillium dangeardii Pitt, Genus Penicillium: 472. 1980 ["1979"]. – Holotype: No. 197477 (IMI).

Penicillium decumbens Thom in U.S.D.A. Bur. Anim. Industr. Bull. 118: 71. 1910. – Neotype (Pitt, 1980): No. 190875 (IMI).

Penicillium dendriticum Pitt, Genus Penicillium: 413. 1980

["1979"]. – Holotype: No. 216897 (IMI).

Penicillium derxii Takada & Udagawa in Mycotaxon 31: 418. 6 Mai 1988. – Holotype: No. 2980 p.p. (NHL).

Penicillium dierckxii Biourge in Cellule 33: 313. 1923. – Neotype (designated here): No. 92216 (IMI).

Penicillium digitatum (Pers.: Fr.) Sacc., Fung. Ital.: tab. 894. Jul 1881. – Basionym: *Monilia digitata* Pers., Syn. Meth. Fung.: 693. 31 Dec 1801 : Fr., Syst. Mycol. 3(2): 411. 1832. – Lectotype (Pitt, 1980): icon in Saccardo, Fung. Ital.: tab. 894. Jul 1881.

Penicillium dimorphosporum H. Swart in Trans. Brit. Mycol. Soc. 55: 310. 19 Oct 1970. – Holotype: No. 456.70 (CBS).

Penicillium diversum Raper & Fennell in Mycologia 40: 539. Sep-Oct 1948. – Neotype (Pitt, 1980): No. 40579 (IMI).

Penicillium dodgei Pitt, Genus Penicillium: 117. 1980 ["1979"]. – Holotype: No. 216896 (IMI).

Penicillium donkii Stolk in Persoonia 7: 333. 20 Jul 1973. – Holotype: No. 188.72 (CBS).

Penicillium duclauxii Delacr. in Bull. Soc. Mycol. France 7: 107. 1891. – Neotype (Pitt, 1980): No. 24312 (IMI).

Penicillium dupontii Griffon & Maubl. in Bull. Soc. Mycol. France 27: 73. 1911. – Neotype (Pitt, 1980): No. 236.58 (CBS).

Penicillium echinulatum Raper & Thom ex Fassat. in Acta Univ. Carol., Biol. 1974: 326. Dec 1977. – Holotype: No. 778523 (PRM).

Penicillium emmonsii Pitt, Genus Penicillium: 479. 1980 ["1979"]. – Holotype: No. 39805 p.p. (IMI).

Penicillium erubescens D. B. Scott in Mycopathol. Mycol. Appl. 36: 14. 31 Oct 1968. – Holotype: No. 318.67 p.p. (CBS).

Penicillium erythromellis A. D. Hocking in Pitt, Genus Penicillium: 459. 1980 ["1979"]. – Holotype: No. 216899 (IMI).

Penicillium estinogenum A. Komatsu & S. Abe ex G. Sm. in Trans. Brit. Mycol. Soc. 46: 335. 21 Oct 1963. – Neotype (designated here): No. 68241 (IMI).

Penicillium expansum Link, in Ges. Naturf. Freunde Berlin Mag. Neuesten Entdeck. Gesammten Naturk. 3: 16. 1809. – Neotype (Samson & al., 1976): No. 325.48 (CBS).

Penicillium fellutanum Biourge in Cellule 33: 262. 1923. – Neotype (Pitt, 1980): No. 39734 (IMI).

Penicillium fennelliae Stolk in Antonie van Leeuwenhoek J. Microbiol. Serol. 35: 261. 1969. – Holotype: No. 711.68 (CBS).

Penicillium flavidostipitatum C. Ramírez & C. C. González in Mycopathologia 88: 3. 1984. – Neotype (Frisvad & al., 1990b): No. 202.87 (CBS).

Penicillium formosanum H. M. Hsieh & al. in Trans. Mycol. Soc.

Republ. China 2: 159. 1987. – Holotype: No. 10001 (PPEH).

Penicillium fractum Udagawa in Trans. Mycol. Soc. Japan 9: 51. 1 Nov 1968. – Holotype: No. 6104 p.p. (NHL).

Penicillium funiculosum Thom in U.S.D.A. Bur. Anim. Industr. Bull. 118: 69. 1910. – Neotype (Pitt, 1980): No. 193019 (IMI).

Penicillium galapagense Samson & Mahoney in Trans. Brit. Mycol. Soc. 69: 158. 19 Aug 1977. – Holotype: No. 751.74 p.p. (CBS).

Penicillium glabrum (Wehmer) Westling in Ark. Bot. 11(1): 131. 1911. – Basionym: *Citromyces glaber* Wehmer, Beitr. Einh. Pilze 1: 24. Jul 1893. – Neotype (Pitt, 1980): No. 91944 (IMI).

Penicillium gladioli L. McCulloch & Thom in Science, ser. 2, 67: 217. 1928. – Neotype (Pitt, 1980): No. 34911 (IMI).

Penicillium glandicola (Oudem.) Seifert & Samson in Samson & Pitt, Adv. Penicillium Aspergillus Syst.: 147. 1985. – Basionym: *Coremiun glandicola* Oudem. in Ned. Kruidk. Arch., ser. 3, 2: 918. Jun 1903. – Holotype: Netherlands, Valkenburg, Jul 1901, *Rick* in herb. Oudemans (L).

Penicillium gracilentum Udagawa & Y. Horie in Trans. Mycol. Soc. Japan 14: 373. 20 Dec 1973. – Holotype: No. 6452 p.p. (NHL).

Penicillium griseofulvum Dierckx in Ann. Soc. Sci. Bruxelles 25: 88. 1901. – Neotype (Pitt, 1980): No. 75832 (IMI).

Penicillium griseopurpureum G. Sm. in Trans. Brit. Mycol. Soc. 48: 275. 22 Jun 1965. – Neotype (designated here): No. 96157 (IMI).

Penicillium herquei Bainier & Sartory in Bull. Soc. Mycol. France 28: 121. 1912. – Neotype (Pitt, 1980): No. 28809 (IMI).

Penicillium heteromorphum H. Z. Kong & Z. T. Qi in Mycosystema 1: 107. 29 Feb 1988. – Neotype (Frisvad & al., 1990b): No. 226.89 (CBS).

Penicillium hirayamae Udagawa in J. Agric. Sci. Tokyo Nogyo Daigaku 5: 6. 1959. – Neotype (Pitt, 1980): No. 78255 (IMI).

Penicillium hirsutum Dierckx in Ann. Soc. Sci. Bruxelles 25: 89. 1901. – Neotype (Pitt, 1980): No. 40213 (IMI).

Penicillium hordei Stolk in Antonie van Leeuwenhoek J. Microbiol. Serol. 35: 270. 1969. – Holotype: No. 701.68 (CBS).

Penicillium implicatum Biourge in Cellule 33: 278. 1923. – Neotype (Pitt, 1980): No. 190235 (IMI).

Penicillium inflatum Stolk & Malla in Persoonia 6: 197. 23 Mar 1971. – Holotype: No. 682.70 (CBS).

Penicillium indonesiae Pitt, Genus Penicillium: 114. 1980 ["1979"]. – Holotype: No. 39733 (IMI).

Penicillium inusitatum D. B. Scott in Mycopathol. Mycol. Appl. 36: 20. 31 Oct 1968. – Holotype: No. 351.67 p.p. (CBS).

Penicillium isariiforme Stolk & J. Mey. in Trans. Brit. Mycol. Soc. 40: 187. 1 Jul 1957. – Neotype (Pitt, 1980): No. 60371 (IMI).

Penicillium islandicum Sopp in Skr. Vidensk.-Selsk. Christiana, Math.-Naturvidensk. Kl. 11: 161. 1912. – Neotype (Pitt, 1980): No. 40042 (IMI).

Penicillium italicum Wehmer in Hedwigia 33: 211. 1 Aug 1894. – Neotype (Samson & al., 1976): No. 339.48 (CBS).

Penicillium janczewskii K. M. Zalessky in Bull. Int. Acad. Polon. Sci., Cl. Sci. Math., Sér. B, Sci. Nat., 1927: 488. 1927. – Neotype (Pitt, 1980): No. 191499 (IMI).

Penicillium janthinellum Biourge in Cellule 33: 258. 1923. – Neotype (Pitt, 1980): No. 40238 (IMI).

Penicillium jensenii K. M. Zalessky in Bull. Int. Acad. Polon. Sci., Cl. Sci. Math., Sér. B, Sci. Nat., 1927: 494. 1927. – Neotype (Pitt, 1980): No. 39768 (IMI).

Penicillium jugoslavicum C. Ramírez & Munt.-Cvetk. in Mycopathologia 88: 65. 1984. – Neotype (Frisvad & al., 1990b): No. 192.87 (CBS).

Penicillium kabunicum Baghd. in Novosti Sist. Nizš. Rast. 1968: 98. 16 Oct 1968. – Neotype (designated here): No. 409.69 (CBS).

Penicillium katangense Stolk in Antonie van Leeuwenhoek J. Microbiol. Serol. 34: 42. 1968. – Holotype: No. 247.67 p.p. (CBS).

Penicillium klebahnii Pitt, Genus Penicillium: 122. 1980 ["1979"]. – Holotype: No. 39737 (IMI).

Penicillium kloeckeri Pitt, Genus Penicillium: 491. 1980 ["1979"]. – Holotype (Pitt, 1980): No. 40047 p.p. (IMI).

Penicillium lanosum Westling in Ark. Bot. 11(1): 97. 1911. – Neotype (designated here): No. 40224 (IMI).

Penicillium lapatayae C. Ramírez in Mycopathologia 91: 96. 1985. – Neotype (Frisvad & al., 1990b): No. 203.87 (CBS).

Penicillium lapidosum Raper & Fennell in Mycologia 40: 524. Sep-Oct 1948. – Neotype (Pitt, 1980): No. 39743 (IMI).

Penicillium lassenii Paden in Mycopathol. Mycol. Appl. 43: 266. 25 Mar 1971. – Holotype: No. JWP 69-26 (UVIC).

Penicillium lehmanii Pitt, Genus Penicillium: 497. 1980 ["1979"]. – Holotype: No. 40043 (IMI).

Penicillium lignorum Stolk in Antonie van Leeuwenhoek J. Microbiol. Serol. 35: 264. 1969. – Holotype: No. 709.68 (CBS).

Penicillium lineatum Pitt, Genus Penicillium: 485. 1980 ["1979"]. – Holotype: No. 39741 p.p. (IMI).

Penicillium lineolatum Udagawa & Y. Horie in Mycotaxon 5: 493. 6 Mai 1977. – Holotype: No. 2776 p.p. (NHL).

Penicillium lividum Westling in Ark. Bot. 11(1): 134. 1911. –

Neotype (Pitt, 1980): No. 39736 (IMI).

Penicillium loliense Pitt, Genus Penicillium: 450. 1980 ["1979"]. – Holotype: No. 216901 (IMI).

Penicillium ludwigii Udagawa in Trans. Mycol. Soc. Japan 10: 2. 1 Aug 1969. – Holotype: No. 6118 p.p. (NHL).

Penicillium maclennaniae H. Y. Yip in Trans. Brit. Mycol. Soc. 77: 202. 6 Aug 1981. – Holotype: No. 35238 (DAR).

Penicillium macrosporum Frisvad & al. in Antonie van Leeuwenhoek J. Microbiol. Serol. 57: 186. 1990. – Holotype: No. 317.63 p.p. (CBS).

Penicillium madriti G. Sm. in Trans. Brit. Mycol. Soc. 44: 44. 21 Mar 1961. – Holotype (Pitt, 1980): No. 86563 (IMI).

Penicillium manginii Duché & R. Heim in Trav. Cryptog. Louis L. Mangin: 450. Sep 1931. – Neotype (designated here): No. 253.31 (CBS).

Penicillium mariaecrucis Quintan. in Avances Nutr. Mejora Anim. Aliment. 23: 334. 1982. – Neotype (Frisvad & al., 1990b): No. 270.83 (CBS).

Penicillium marneffei Segretain & al. in Bull. Soc. Mycol. France 75: 416. 28 Feb 1960. – Neotype (Pitt, 1980): No. 68794iii (IMI).

Penicillium megasporum Orpurt & Fennell in Mycologia 47: 233. 29 Apr 1955. – Neotype (Pitt, 1980): No. 216904 (IMI).

Penicillium melinii Thom, Penicillia: 273. Jan 1930. – Neotype (Pitt, 1980): No. 40216 (IMI).

Penicillium meliforme Udagawa & Y. Horie in Trans. Mycol. Soc. Japan 14: 376. 20 Dec 1973. – Holotype: No. 6468 p.p. (NHL).

Penicillium meridianum D. B. Scott in Mycopathol. Mycol. Appl. 36: 12. 31 Oct 1968. – Holotype: No. 314.67 p.p. (CBS).

Penicillium miczynskii K. M. Zalessky in Bull. Int. Acad. Polon. Sci., Cl. Sci. Math., Sér. B, Sci. Nat., 1927: 482. 1927. – Neotype (Pitt, 1980): No. 40030 (IMI).

Penicillium mimosinum A. D. Hocking in Pitt, Genus Penicillium: 507. 1980 ["1979"]. – Holotype: No. 223991 p.p. (IMI).

Penicillium minioluteum Dierckx in Ann. Soc. Sci. Bruxelles 25: 87. 1901. – Neotype (designated here): No. 642.68 (CBS).

Penicillium mirabile Beliakova & Milko in Mikol. & Fitopatol. 6: 145. 1972. – Holotype: No. F-1328 (BKM).

Penicillium moldavicum Milko & Beliakova in Novosti Sist. Nizš. Rast. 1967: 255. 28 Feb 1967. – Neotype (designated here): No. 129966 (IMI).

Penicillium molle Pitt, Genus Penicillium: 148. 1980 ["1979"]. – Holotype: No. 84589 (IMI).

Penicillium mononematosum (Frisvad & al.) Frisvad in Mycologia 81: 857. 11 Jan 1990. – Basionym: *Penicillium glandicola* var. *mononematosum* Frisvad & al. in

Canad. J. Bot. 65: 767. 23 Apr 1987. – Holotype (Frisvad & al., 1987): No. 296925 (IMI).

Penicillium montanense M. Chr. & Backus in Mycologia 54: 574. 25 Jan 1963. – Holotype: Cryptogamic Herb. No. GW1-6 (WIS).

Penicillium nalgiovense Laxa in Zentralbl. Bakteriol., 2. Abt., 86: 160. 22 Jun 1932. – Neotype (designated here): No. 352.48 (CBS).

Penicillium nepalense Takada & Udagawa in Trans. Mycol. Soc. Japan 24: 146. Jul 1983. – Holotype: No. 6482 p.p. (NHL).

Penicillium nilense Pitt, Genus Penicillium: 145. 1980 ["1979"]. – Holotype: No. 40580 (IMI).

Penicillium nodositanum Valla in Pl. & Soil 114: 146. 1989. – Neotype (designated here): No. 330.90 (CBS).

Penicillium nodulum H. Z. Kong & Z. T. Qi in Mycosystema 1: 108. 29 Feb 1988. – Neotype (Frisvad & al., 1990b): No. 227.89 (CBS).

Penicillium novae-zeelandiae J. F. H. Beyma in Antonie van Leeuwenhoek J. Microbiol. Serol. 6: 275. 1940. – Neotype (Pitt, 1980): No. 40584ii (IMI).

Penicillium ochrochloron Biourge in Cellule 33: 269. 1923. – Neotype (Pitt, 1980): No. 39806 (IMI).

Penicillium ochrosalmoneum Udagawa in J. Agric. Sci. Tokyo Nogyo Daigaku 5: 10. 1959. – Holotype: No. 6048 (NHL).

Penicillium oblatum Pitt & A. D. Hocking in Mycologia 77: 819. 15 Oct 1985. – Holotype: No. 2234 (FRR).

Penicillium ohiense L. H. Huang & J. A. Schmitt in Ohio J. Sci. 75: 78. 1975. – Holotype: No. 6086 p.p. (NHL).

Penicillium olsonii Bainier & Sartory in Ann. Mycol. 10: 398. 10 Aug 1912. – Neotype (Pitt, 1980): No. 192502 (IMI).

Penicillium onobense C. Ramírez & A. T. Martínez in Mycopathologia 74: 44. 1981. – Neotype (Frisvad & al., 1990b): No. 174.81 (CBS).

Penicillium ornatum Udagawa in Trans. Mycol. Soc. Japan 9: 49. 1 Nov 1968. – Holotype: No. 6101 p.p. (NHL).

Penicillium osmophilum Stolk & Vccnb.-Rijks in Antonie van Leeuwenhoek J. Microbiol. Serol. 40: 1. 1974. – Holotype: No. 462.72 p.p. (CBS).

Penicillium oxalicum Currie & Thom in J. Biol. Chem. 22: 289. 1915. – Neotype (Pitt, 1980): No. 192332 (IMI).

Penicillium palmae Samson & al. in Stud. Mycol. 31: 135. 1 Jul 1989. – Holotype: No. 442.88 (CBS).

Penicillium palmense C. Ramírez & al. in Mycopathologia 66: 80. 29 Dec 1978. – Neotype (Frisvad & al., 1990b): No. 336.79 (CBS).

Penicillium panamense Samson & al. in Stud. Mycol. 31: 136. 1 Jul

1989. – Holotype: No. 128.89 (CBS).

Penicillium panasenkoi Pitt, Genus Penicillium: 482. 1980 ["1979"]. – Holotype: No. 129962 (IMI).

Penicillium papuanum Udagawa & Y. Horie in Trans. Mycol. Soc. Japan 14: 378. 20 Dec 1973. – Holotype: No. 6463 (NHL).

Penicillium paraherquei S. Abe ex G. Sm. in Trans. Brit. Mycol. Soc. 46: 335. 21 Oct 1963. – Neotype (designated here): No. 68220 (IMI).

Penicillium patens Pitt & A. D. Hocking in Mycotaxon 22: 205. 8 Feb 1985. – Holotype: No. 2661 (FRR).

Penicillium paxilli Bainier in Bull. Soc. Mycol. France 23: 95. 1907. – Neotype (Pitt, 1980): No. 40226 (IMI).

Penicillium pedemontanum Mosca & A. Fontana in Allionia 9: 40. 1963. – Neotype (designated here): No. 265.65 (CBS).

Penicillium phoeniceum J. F. H. Beyma in Zentralbl. Bakteriol., 2. Abt., 88: 136. 24 Apr 1933. – Neotype (Pitt, 1980): No. 40585 (IMI).

Penicillium piceum Raper & Fennell in Mycologia 40: 533. Sep-Oct 1948. – Neotype (Pitt, 1980): No. 40038 (IMI).

Penicillium pinetorum M. Chr. & Backus in Mycologia 53: 457. 21 Jun 1962. – Holotype: No. WSF 15-C (WIS).

Penicillium pinophilum Hedgc. in U.S.D.A. Bur. Anim. Industr. Bull. 118: 37. 1910. – Neotype (Pitt, 1980): No. 114933 (IMI).

Penicillium piscarium Westling in Ark. Bot. 11(1): 86. 1911. – Neotype (designated here): No. 40032 (IMI).

Penicillium pittii Quintan. in Mycopathologia 91: 75. 1985. – Neotype (Frisvad & al., 1990b): No. 139.84 (CBS).

Penicillium primulinum Pitt, Genus Penicillium: 455. 1980 ["1979"]. – Holotype: No. 40031 (IMI).

Penicillium proteolyticum Kamyschko in Bot. Mater. Otd. Sporov. Rast. 14: 228. 29 Mar 1961. – Neotype (designated here): No. 303.67 (CBS).

Penicillium pseudostromaticum Hodges & al. in Mycologia 62: 1106. 10 Feb 1971. – Holotype: Warner 18 (NY).

Penicillium pulvillorum Turfitt in Trans. Brit. Mycol. Soc. 23: 186. 31 Jul 1939. – Neotype (designated here): No. 280.39 (CBS).

Penicillium purpurascens (Sopp) Biourge in Cellule 33: 105. 1923. – Basionym: *Citromyces purpurascens* Sopp in Skr. Vidensk.-Selsk. Christiana, Math.-Naturvidensk. Kl. 11: 117. 1912. – Neotype (Pitt, 1980): No. 39745 (IMI).

Penicillium purpureum Stolk & Samson in Stud. Mycol. 2: 57. 1 Nov 1972. – Lectotype (Pitt,

1980): icon in Stud. Mycol. 2: 59. 1 Nov 1972.

Penicillium purpurogenum Stoll, Beitr. Charakt. Penicill.: 32. 1904. – Neotype (Pitt, 1980): No. 91926 (IMI).

Penicillium raciborskii K. M. Zalessky in Bull. Int. Acad. Polon. Sci., Cl. Sci. Math., Sér. B, Sci. Nat., 1927: 454. 1927. – Neotype (designated here): No. 40568 (IMI).

Penicillium rademiricii Quintan. in Mycopathologia 91: 72. 1985. – Neotype (Frisvad & al., 1990b): No. 140.84 (CBS).

Penicillium raperi G. Sm. in Trans. Brit. Mycol. Soc. 40: 486. 20 Dec 1957. – Neotype (designated here): No. 71625 (IMI).

Penicillium raistrickii G. Sm. in Trans. Brit. Mycol. Soc. 18: 90. 16 Aug 1933. – Neotype (Pitt, 1980): No. 40221 (IMI).

Penicillium rasile Pitt, Genus Penicillium: 120. 1980 ["1979"]. – Holotype: No. 39735 (IMI).

Penicillium resedanum McLennan & Ducker in Austral. J. Bot. 2: 360. Nov 1954. – Neotype (Pitt, 1980): No. 62877 (IMI).

Penicillium restrictum J. C. Gilman & E. V. Abbott in Iowa State Coll. J. Sci. 1: 297. 1 Apr 1927. – Neotype (Pitt, 1980): No. 40228 (IMI).

Penicillium reticulisporum Udagawa in Trans. Mycol. Soc. Japan 9: 52. 1 Nov 1968. – Holotype: No. 6105 p.p. (NHL).

Penicillium rolfsii Thom, Penicillia: 489. Jan 1930. – Neotype (Pitt, 1980): No. 40029 (IMI).

Penicillium roqueforti Thom in U.S.D.A. Bur. Anim. Industr. Bull. 82: 35. 1906. – Neotype (Pitt, 1980): No. 24313 (IMI).

Penicillium roseopurpureum Dierckx in Ann. Soc. Sci. Bruxelles 25: 86. 1901. – Neotype (Pitt, 1980): No. 40573 (IMI).

Penicillium rubefaciens Quintan. in Mycopathologia 80: 73. 1982. – Neotype (Frisvad & al., 1990b): No. 145.83 (CBS).

Penicillium rubidurum Udagawa & Y. Horie in Trans. Mycol. Soc. Japan 14: 381. 20 Dec 1973. – Holotype: No. 6460 p.p. (NHL).

Penicillium rugulosum Thom in U.S.D.A. Bur. Anim. Industr. Bull. 118: 60. 1910. – Neotype (Pitt, 1980): No. 40041 (IMI).

Penicillium sabulosum Pitt & A. D. Hocking in Mycologia 77: 818. 15 Oct 1985. – Holotype: No. 2743 (FRR).

Penicillium sacculum E. Dale in Ann. Mycol. 24: 137. 20 Jun 1926. – Neotype (designated here): No. 231.61 (CBS).

Penicillium sajarovii Quintan. in Avances Nutr. Mejora Anim. Aliment. 22: 539. 1981. – Neotype (Frisvad & al., 1990b): No. 277.83 (CBS).

Penicillium scabrosum Frisvad & al. in Persoonia 14: 177. 27 Jun 1990. – Holotype: No. 285533 (IMI).

Penicillium sclerotigenum W. Yamam. in Sci. Rep. Hyogo Univ. Agric., Ser. Agric. Biol., ser. 2, 1: 69. 1955. – Neotype (Pitt, 1980): No. 68616 (IMI).

Penicillium sclerotiorum J. F. H. Beyma in Zentralbl. Bakteriol., 2. Abt., 96: 418. 10 Aug 1937. – Neotype (Pitt, 1980): No. 40569 (IMI).

Penicillium senticosum D. B. Scott in Mycopathol. Mycol. Appl. 36: 5. 31 Oct 1968. – Holotype: No. 316.67 p.p. (CBS).

Penicillium shearii Stolk & D. B. Scott in Persoonia 4: 396. 1 Aug 1967. – Holotype: No. 290.48 p.p. (CBS).

Penicillium shennangjianum H. Z. Kong & Z. T. Qi in Mycosystema 1: 110. 29 Feb 1988. – Neotype (Frisvad & al., 1990b): No. 228.89 (CBS).

Penicillium siamense Manoch & C. Ramírez in Mycopathologia 101: 32. Jan 1988. – Neotype (designated here): No. 475.88 (CBS).

Penicillium simplicissimum (Oudem.) Thom, Penicillia: 335. Jan 1930. – Basionym: *Spicaria simplicissima* Oudem. in Ned. Kruidk. Arch., ser. 3, 2: 763. Jun 1902. – Neotype (Jensen, 1912): No. 5921 (CUP).

Penicillium sinaicum Udagawa & S. Ueda in Mycotaxon 14: 266. 22 Jan 1982. – Holotype: No. 2894 p.p. (NHL).

Penicillium sizovae Baghd. in Novosti Sist. Nizš. Rast. 1968: 103.

16 Oct 1968. – Neotype (designated here): No. 413.69 (CBS).

Penicillium skrjabinii Schmotina & Golovleva in Mikol. & Fitopatol. 8: 530. 1974. – Neotype (designated here): No. 196528 (IMI).

Penicillium smithii Quintan. in Avances Nutr. Mejora Anim. Aliment. 23: 340. 1982. – Neotype (designated here): No. 276.83 (CBS).

Penicillium solitum Westling in Ark. Bot. 11(1): 65. 1911. – Neotype (Frisvad & al., 1990a): No. 424.89 (CBS).

Penicillium soppii K. M. Zalessky in Bull. Int. Acad. Polon. Sci., Cl. Sci. Math., Sér. B, Sci. Nat., 1927: 476. 1927. – Neotype (designated here): No. 40217 (IMI).

Penicillium sphaerum Pitt, Genus Penicillium: 494. 1980 ["1979"]. – Holotype: No. 40589 p.p. (IMI).

Penicillium spinulosum Thom in U.S.D.A. Bur. Anim. Industr. Bull. 118: 76. 1910. – Neotype (Pitt, 1980): No. 24316i (IMI).

Penicillium spirillum Pitt, Genus Penicillium: 476. 1980 ["1979"]. – Holotype: No. 40593 p.p. (IMI).

Penicillium steckii K. M. Zalessky in Bull. Int. Acad. Polon. Sci., Cl. Sci. Math., Sér. B, Sci. Nat., 1927: 469. 1927. – Neotype (designated here): No. 40583 (IMI).

Penicillium stolkiae D. B. Scott in Mycopathol. Mycol. Appl. 36: 8. 31 Oct 1968. – Holotype: No. 315.67 p.p. (CBS).

Penicillium striatisporum Stolk in Antonie van Leeuwenhoek J. Microbiol. Serol. 35: 268. 1969. – Holotype: No. 705.68 (CBS).

Penicillium sublateritium Biourge in Cellule 33: 315. 1923. – Neotype (Pitt, 1980): No. 40594 (IMI).

Penicillium syriacum Baghd. in Novosti Sist. Nizš. Rast. 1968: 111. 16 Oct 1968. – Neotype (designated here): No. 418.69 (CBS).

Penicillium tardum Thom, Penicillia: 485. Jan 1930. – Neotype (designated here): No. 40034 (IMI).

Penicillium terlikowskii K. M. Zalessky in Bull. Int. Acad. Polon. Sci., Cl. Sci. Math., Sér. B, Sci. Nat., 1927: 501. 1927. – Neotype (designated here): No. 228.28 (CBS).

Penicillium terrenum D. B. Scott in Mycopathol. Mycol. Appl. 36: 1. 31 Oct 1968. – Holotype: No. 313.67 p.p. (CBS).

Penicillium thomii Maire in Bull. Soc. Hist. Nat. Afrique N. 8: 189. 1917. – Neotype (Pitt, 1980): No. 189694 (IMI).

Penicillium tularense Paden in Mycopathol. Mycol. Appl. 43: 264. 25 Mar 1971. – Holotype: No. JWP 68-31 p.p. (UVIC).

Penicillium turbatum Westling in Ark. Bot. 11(1): 128. 1911. – Neotype (Pitt, 1980): No. 39738 (IMI).

Penicillium udagawae Stolk & Samson in Stud. Mycol. 2: 36. 1 Nov 1972. – Holotype: No. 579.72 p.p. (CBS).

Penicillium ulaiense H. M. Hsieh & al. in Trans. Mycol. Soc. Republ. China 2: 161. 1987. – Holotype: No. 29001.87 (PPEH)

Penicillium unicum Tzean & al. in Mycologia 84: 739. 21 Oct 1992. – Holotype: No. PPH16 p.p. (PPEH).

Penicillium vanbeymae Pitt, Genus Penicillium: 142. 1980 ["1979"]. – Holotype: No. 40590 (IMI).

Penicillium variabile Sopp in Skr. Vidensk.-Selsk. Christiana, Math.-Naturvidensk. Kl. 11: 169. 1912. – Neotype (Pitt, 1980): No. 40040 (IMI).

Penicillium varians G. Sm. in Trans. Brit. Mycol. Soc. 18: 89. 16 Aug 1933. – Neotype (designated here): No. 40586 (IMI).

Penicillium vasconiae C. Ramírez & A. T. Martínez in Mycopathologia 72: 189. 28 Nov 1980. – Neotype (Frisvad & al., 1990b): No. 339.79 (CBS).

Penicillium velutinum J. F. H. Beyma in Zentralbl. Bakteriol., 2. Abt., 91: 353. 28 Feb 1935. – Neotype (Pitt, 1980): No. 40571 (IMI).

Penicillium verrucosum Dierckx in Ann. Soc. Sci. Bruxelles 25: 88. 1901. – Neotype (Pitt, 1980): No. 200310 (IMI).

Penicillium verruculosum Peyronel, Germi Atmosf. Fung. Micel.: 22. 1913. – Neotype (Pitt, 1980): No. 40039 (IMI).

Penicillium vinaceum J. C. Gilman & E. V. Abbott in Iowa State Coll. J. Sci. 1: 299. 1 Apr 1927. – Neotype (Pitt, 1980): No. 29189 (IMI).

Penicillium viridicatum Westling in Ark. Bot. 11(1): 88. 1911. – Neotype (Pitt, 1980): No. 39758ii (IMI).

Penicillium vulpinum (Cooke & Massee) Seifert & Samson in Samson & Pitt, Adv. Penicillium Aspergillus Syst.: 144. 1985. – Basionym: *Coremium vulpinum* Cooke & Massee in Grevillea 16: 81. Mar 1888. – Holotype (see Seifert & Samson, 1985): "on dung", *s.coll.,* in herb. Cooke (K).

Penicillium waksmanii K. M. Zalessky in Bull. Int. Acad. Polon. Sci., Cl. Sci. Math., Sér. B, Sci. Nat., 1927: 468. 1927. – Neotype (Pitt, 1980): No. 39746i (IMI).

Penicillium westlingii K. M. Zalessky in Bull. Int. Acad. Polon. Sci., Cl. Sci. Math., Sér. B, Sci. Nat., 1927: 473. 1927. – Neotype (designated here): No. 92272 (IMI).

Penicillium zonatum Hodges & J. J. Perry in Mycologia 65: 697. 19 Jul 1973. – Holotype: No. FSL 525 p.p. (BPI).

Petromyces Malloch & Cain
(holomorphs)

Petromyces albertensis J. P. Tewari in Mycologia 77: 114. 15 Feb 1985. – Holotype: No. 2976 (UAMH). [Anamorph: *Aspergillus albertensis* J. P. Tewari].

Petromyces alliaceus Malloch & Cain in Canad. J. Bot. 50: 2623. 26 Jan 1973. – Holotype: No. 46232 (TRTC). [Anamorph: *Aspergillus alliaceus* Thom & Church].

Sarophorum Syd. & P. Syd.
(anamorphs)

Sarophorum palmicola (Henn.) Seifert & Samson in Samson & Pitt, Adv. Penicillium Aspergillus Syst.: 403. 1985. – Basionym: *Penicilliopsis palmicola* Henn. in Hedwigia 43: 352. 15 Jul 1904. – Lectotype (designated here; see also Seifert & Samson, 1985): Brazil, Jurna, Jul 1901, *Ule 2834* (L).

Sclerocleista Subram.
(holomorphs)

Sclerocleista ornata (Raper & al.) Subram. in Curr. Sci. 41: 757. 5 Nov 1972. – Basionym: *Aspergillus ornatus* Raper & al. in Mycologia 45: 678. 9 Oct 1953 (nom. holomorph.). – Neotype (Samson & Gams, 1985): No. 55295 (IMI). [Anamorph: *Aspergillus ornatulus* Samson & W. Gams].

Sclerocleista thaxteri Subram. in Curr. Sci. 41: 757. 5 Nov 1972. – Holotype (Subramanian, 1972): ex caterpillar dung, Kittery Point, *R. Thaxter* (FH). [Anamorph: -*Aspergillus citrisporus* Höhn.].

Stilbodendron Syd. & P. Syd.
(anamorphs)

Stilbodendron cervinum (Cooke & Massee) Samson & Seifert in Samson & Pitt, Adv. Penicillium Aspergillus Syst.: 408. 1985. – Basionym: *Corallodendron cervinum* Cooke & Massee in Grevillea 16: 71. Mar 1888. – Holotype (see Samson & Seifert, 1985): Africa, *Holmes* (K).

Talaromyces C. R. Benj.
(holomorphs)

Talaromyces assiutensis Samson & Abdel-Fattah in Persoonia 9: 501. 13 Jul 1978. – Holotype: No. 147.78 p.p. (CBS). [Anamorph: *Penicillium assiutense* Samson & Abdel-Fattah].

Talaromyces avellaneus (Thom & Turesson) C. R. Benj. in Mycologia 47: 682. 7 Oct 1955. – Basionym: *Penicillium avellaneum* Thom & Turesson in Mycologia 7: 284. Sep 1915 (nom. holomorph.). – Neotype (designated here): No. 40230 (IMI). [Anamorph: *Merimbla ingelheimense* (J. F. H. Beyma) Pitt].

Talaromyces bacillisporus (Swift) C. R. Benj. in Mycologia 47: 684. 7 Oct 1955. – Basionym: *Penicillium bacillisporum* Swift in Bull. Torrey Bot. Club 59: 221. 4 Mai 1932 (nom. holomorph.). – Holotype (Stolk & Samson, 1972): No. 296.48 (CBS). [Anamorph: *Geosmithia swiftii* Pitt].

Talaromyces byssochlamydoides Stolk & Samson in Stud. Mycol. 2: 45. 1 Nov 1972. – Holotype: No. 413.74 (CBS). [Anamorph: *Paecilomyces byssochlamydoides* Stolk & Samson].

Talaromyces derxii Takada & Udagawa in Mycotaxon 31: 418. 6 Mai 1988. – Holotype: No. 2980 p.p. (NHL). [Anamorph: *Penicillium derxii* Takada & Udagawa].

Talaromyces emersonii Stolk in Antonie van Leeuwenhoek J. Microbiol. Serol. 31: 262. 1965. – Holotype: No. 393.64 p.p. (CBS). [Anamorph: *Geosmithia emersonii* (Stolk) Pitt].

Talaromyces flavus (Klöcker) Stolk & Samson in Stud. Mycol. 2: 10. 1 Nov 1972. – Basionym: *Gymnoascus flavus* Klöcker in Hedwigia 41: 80. 24 Apr 1902. – Neotype (Stolk & Samson, 1972): No. 310.38 p.p. (CBS). [Anamorph: *Penicillium dangeardii* Pitt].

Talaromyces galapagensis Samson & Mahoney in Trans. Brit. Mycol. Soc. 69: 158. 19 Aug 1977. – Holotype: No. 751.74 p.p. (CBS). [Anamorph: *Penicillium galapagense* Samson & Mahoney].

Talaromyces helicus (Raper & Fennell) C. R. Benj. in Mycologia 47: 684. 7 Oct 1955. – Basionym: *Penicillium helicum* Raper & Fennell in Mycologia 40: 515. Sep-Oct 1948 (nom. holomorph.). – Neotype (Pitt, 1980): No. 40593 p.p. (IMI). [Anamorph: *Penicillium spirillum* Pitt].

Talaromyces leycettanus H. C. Evans & Stolk in Trans. Brit.

Mycol. Soc. 56: 45. 29 Feb 1971. – Holotype: No. 398.68 (CBS). [Anamorph: *Paecilomyces leycettanus* (H. C. Evans & Stolk) Stolk & al.].

Talaromyces luteus (Zukal) C. R. Benj. in Mycologia 47: 681. 7 Oct 1955. – Basionym: *Penicillium luteum* Zukal in Sitzungsber. Kaiserl. Akad. Wiss., Math.-Naturwiss. Cl., Abt. 1, 98: 561. 1890 ["1890"] (nom. holomorph.). – Neotype (Pitt, 1980): No. 89305 (IMI).

Talaromyces macrosporus (Stolk & Samson) Frisvad & al. in Antonie van Leeuwenhoek J. Microbiol. Serol. 57: 186. 1990. – Basionym: *Talaromyces flavus* var. *macrosporus* Stolk & Samson in Stud. Mycol. 2: 15. 1 Nov 1972. – Holotype (Stolk & Samson, 1972): No. 317.63 p.p. (CBS). [Anamorph: *Penicillium macrosporum* Frisvad & al.].

Talaromyces mimosinus A. D. Hocking in Pitt, Genus Penicillium: 507. 1980 ["1979"]. – Holotype: No. 223991 p.p. (IMI). [Anamorph: *Penicillium mimosinum* A. D. Hocking].

Talaromyces ohiensis Pitt, Genus Penicillium: 502. 1980 ["1979"]. – Holotype: No. 6086 p.p. (NHL). [Anamorph: *Penicillium ohiense* L. H. Huang & J. A. Schmitt].

Talaromyces panasenkoi Pitt, Genus Penicillium: 482. 1980 ["1979"]. – Holotype: icon of 'Penicillium ucrainicum' in Mycologia 56: 59. 20 Feb 1964. [Anamorph: *Penicillium panasenkoi* Pitt].

Talaromyces purpureus (E. Müll. & Pacha-Aue) Stolk & Samson in Stud. Mycol. 2: 57. 1 Nov 1972. – Basionym: *Arachniotus purpureus* E. Müll. & Pacha-Aue in Nova Hedwigia 15: 552. 5 Dec 1968. – Lectotype (Pitt, 1980): icon in Stud. Mycol. 2: 59. 1 Nov 1972. [Anamorph: *Penicillium purpureum* Stolk & Samson [Anamorph: *Penicillium sphaerum* Pitt].

Talaromyces rotundus (Raper & Fennell) C. R. Benj. in Mycologia 47: 683. 7 Oct 1955. – Basionym: *Penicillium rotundum* Raper & Fennell in Mycologia 40: 518. Sep-Oct 1948 (nom. holomorph.). – Neotype (Pitt, 1980): No. 40589 p.p. (IMI).

Talaromyces stipitatus (Thom) C. R. Benj. in Mycologia 47: 684. 7 Oct 1955. – Basionym: *Penicillium stipitatum* Thom in Mycologia 27: 138. Mar-Apr 1935 (nom. holomorph.). – Neotype (Pitt, 1980): No. 39805 p.p. (IMI). [Anamorph: *Penicillium emmonsii* Pitt].

Talaromyces striatus (Raper & Fennell) C. R. Benj. in Mycologia 47: 682. 7 Oct 1955. – Basionym: *Penicillium striatum* Raper & Fennell in Mycologia 40: 521. Sep-Oct 1948 (nom. holomorph.). – Neotype (Pitt, 1980): No. 39741 p.p. (IMI). [Anamorph: *Penicillium lineatum* Pitt].

Talaromyces thermophilus Stolk in Antonie van Leeuwenhoek J.

Microbiol. Serol. 31: 268. 1965. –
Holotype: No. 236.58 (CBS).
[Anamorph: *Penicillium dupontii*
Griffon & Maubl.].

Talaromyces trachyspermus (Shear)
Stolk & Samson in Stud. Mycol.
2: 32. 1 Nov 1972. – Basionym:
Arachniotus trachyspermus Shear
in Science, ser. 2, 16: 138. 1902. –
Holotype (Shear, 1902): No. 5798
(BPI). [Anamorph: *Penicillium
lehmanii* Pitt].

Talaromyces udagawae Stolk &
Samson in Stud. Mycol. 2: 36. 1
Nov 1972. – Holotype: No.
579.72 p.p. (CBS). [Anamorph:
Penicillium udagawae Stolk &
Samson].

Talaromyces unicus Tzean & al. in
Mycologia 84: 739. 21 Oct 1992.
– Holotype: No. PPH16 p.p.
(PPEH). [Anamorph: *Penicillium
unicum* Tzean & al.].

Talaromyces wortmannii (Klöcker) C. R. Benj. in Mycologia 47:
683. 7 Oct 1955. – Basionym:
Penicillium wortmannii Klöcker
in Compt.-Rend. Trav. Carlsberg
Lab. 6: 100. 1903 (nom. holomorph.). – Neotype (Pitt, 1980):
No. 40047 p.p. (IMI). [Ana-
morph: *Penicillium kloeckeri*
Pitt].

***Thermoascus* Miehe**
(holomorphs)

Thermoascus aurantiacus Miehe,
Selbsterhitzung Heus: 70. 1907. –
Neotype (Cooney & Emerson,
1964): No. M206516 (UC).

Thermoascus aegyptiacus S. Ueda
& Udagawa in Trans. Mycol. Soc.
Japan 24: 135. Jul 1983. – Holo-
type: No. 2914 (NHL). [Ana-
morph: *Paecilomyces aegyptiacus*
S. Ueda & Udagawa].

Thermoascus crustaceus (Apinis &
Chesters) Stolk in Antonie van
Leeuwenhoek J. Microbiol. Serol.
31: 272. 1965. – Basionym: *Dactylomyces crustaceus* Apinis &
Chesters in Trans. Brit. Mycol.
Soc. 47: 428. 23 Sep 1964. – Neotype (designated here): Herb.
102470 (IMI). [Anamorph: *Paecilomyces crustaceus* Apinis &
Chesters].

Thermoascus thermophilus (Sopp)
Arx, Gen. Fungi Sporul. Pure
Cult., ed. 2: 94. 1974. – Basio-
nym: *Dactylomyces thermophilus*
Sopp in Skr. Vidensk.-Selsk.
Christiana, Math.-Naturvidensk.
Kl. 11: 35. 1912. – Neotype (des-
ignated here): No. 528.71 (CBS).

***Torulomyces* Delitsch**
(anamorphs)

Torulomyces lagena Delitsch, Syst.
Schimmelpilze: 91. 1943. – Neotype (Stolk & Samson, 1983): No.
185.65 (CBS).

***Trichocoma* Jungh.**
(holomorphs)

Trichocoma paradoxa Jungh.,
Praem. Fl. Crypt. Java 1: 9. 1838.
– Holotype: "*Trichocoma para-
doxa*", *Junghuhn* (BO).

Warcupiella Subram.
(holomorphs)

Warcupiella spinulosa (Warcup)
Subram. in Curr. Sci. 41: 757. 5
Nov 1972. – Basionym: _Asper-_

gillus spinulosus Warcup in Raper
& Fennell, Genus Aspergillus:
204. 1965 (nom. holomorph.). –
Neotype (Samson & Gams,
1985): No. 75885 (IMI). [Ana-
morph: _Aspergillus   warcupii_
Samson & W. Gams].

References

Al-Musallam, A. 1980. _Revision of the black_ Aspergillus _species._ Utrecht.

Blaser, P. 1975. Taxonomische und physiologische Untersuchungen über
die Gattung _Eurotium_ Link ex Fries. _Sydowia_ 28: 1-49.

Christensen, M. & Fennell, D. I. 1964. The rediscovery of _Aspergillus
cervinus. Mycologia_ 56: 350-361.

Cooney, D. G. & Emerson, R. 1964. _Thermophilic fungi._ San Francisco.

Eriksson, O. E. & Hawksworth, D. L. 1991. Outline of the Ascomycetes –
1990. _Syst. Ascomycetum_ 9: 39-271.

Frisvad, J. C., Filtenborg, O. & Wicklow, D. T. 1987. Terverticillate penicillia isolated from underground seed caches and cheek pouches of
banner-tailed kangaroo rats _(Dipodomys spectabilis). Canad. J. Bot._ 65:
765-773.

– , Samson, R. A. & Stolk, A. C. 1990a. Notes on the typification of
some species of _Penicillium . Persoonia_ 14: 193-202.

– , Samson, R. A. & Stolk, A. C. 1990b. Disposition of recently described species of _Penicillium . Persoonia_ 14: 209-232.

Holmgren, P. K., Holmgren, N. H. & Barnett, L. C. 1990. Index herbariorum. Part. I: the herbaria of the world. Eighth edition. _Regnum Veg._ 120.

Hughes, S. J. 1951. Studies of microfungi XI. Some Hyphomycetes which
produce phialides. _Mycol. Pap._ 45: 1-36.

Jensen, C. N. 1912. Fungous flora of the soil. _Cornell Univ. Agric. Exp.
Sta. Bull._ 315: 415-501.

Kozakiewicz, Z. 1982. The identity and typification of _Aspergillus parasiticus. Mycotaxon_ 15: 293-305.

– 1989. _Aspergillus_ species on stored products. _Mycol. Pap._ 161: 1-188.

– , Frisvad, J. C., Hawksworth, D. L., Pitt, J. I., Samson, R. A. & Stolk, A.
C. 1992. Proposals for nomina specifica conservanda and rejicienda in
Aspergillus and _Penicillium_ (fungi). _Taxon_ 41: 109-113.

Malloch, D. & Cain, R. F. 1972. The _Trichocomataceae:_ ascomycetes
with _Aspergillus, Paecilomyces_ and _Penicillium_ imperfect states. _Canad.
J. Bot._ 50: 2613-2628.

Pitt, J. I. 1979. *Geosmithia,* gen. nov. for *Penicillium lavendulum* and related species. *Canad. J. Bot.* 57: 2021-2030.

– 1980. The genus *Penicillium* and its teleomorphic states *Eupenicillium* and *Talaromyces.* London.

– & Hocking, A. D. 1979. *Merimbla* gen. nov. for the anamorphic state of *Talaromyces avellaneus. Canad. J. Bot.* 57: 2394-2398.

Samson, R. A. 1974. *Paecilomyces* and some allied Hyphomycetes. *Stud. Mycol.* 6: 1-119.

– & Gams, W. 1985. Typification of the species of *Aspergillus* and associated teleomorphs. Pp. 31-54 *in:* Samson, R. A. & Pitt, J. I. (ed.) *Advances in* Penicillium *and* Aspergillus *Systematics.* New York.

– & Seifert, K. A. 1985. The ascomycete genus *Penicilliopsis* and its anamorphs. Pp. 397-428 *in:* Samson, R. A. & Pitt, J. I. (ed.) *Advances in* Penicillium *and* Aspergillus *Systematics.* New York.

– , Stolk, A. C. & Hadlok, R. 1976. Revision of the Subsection *Fasciculata* of *Penicillium* and some allied species. *Stud. Mycol.* 11: 1-47.

Seifert, K. A. & Samson, R. A. 1985. The genus *Coremium* and the synnematous penicillia. Pp. 143-154 *in:* Samson, R. A. & Pitt, J. I. (ed.) *Advances in* Penicillium *and* Aspergillus *Systematics.* New York.

Shear, C. L. 1902. *Arachniotus trachyspermus,* a new species of the Gymnoascaceae. *Science,* ser. 2, 16: 138.

Stolk, A. C. 1965. Thermophilic species of *Talaromyces* Benjamin and *Thermoascus* Miehe. *Antonie van Leeuwenhoek J. Microbiol. Serol.* 31: 262-276.

– & Samson, R. A. 1972. The genus *Talaromyces.* Studies on *Talaromyces* and related genera II. *Stud. Mycol.* 2: 1-65.

– & – 1983. The Ascomycete genus *Eupenicillium* and related *Penicillium* anamorphs. *Stud. Mycol.* 23: 1-147.

– & Scott, D. B. 1967. Studies on the genus *Eupenicillium* Ludwig. I. Taxonomy and nomenclature of Penicillia in relation to their sclerotioid ascocarpic states. *Persoonia* 4: 391-405.

Subramanian, C. V. 1972. The perfect states of *Aspergillus. Curr. Sci.* 41: 75-761.

Takishima, Y., Shimura, J., Udagawa, Y. & Sugawara, A. 1989. *Guide to worlddata center on microorganisms, with a list of culture collections in the world.* Saitama (Japan).

NAMES IN CURRENT USE IN THE
CLADONIACEAE (LICHEN-FORMING ASCOMYCETES) IN THE RANKS OF GENUS TO VARIETY

By Teuvo Ahti

The present list of names in current use (NCU) in the family *Cladoniaceae* introduces lichens among the groups for which lists down to the levels of species, subspecies and variety have been prepared. The *Cladoniaceae* are a worldwide distributed family with numerous very common species. As to nomenclature, it is a notoriously critical group with some 500 taxa currently recognized but with more than 3000 epithets published. Up to recent times hundreds of infraspecific taxa have been used by some authors, but few of them are currently recognized. The numerous names proposed in the past reflect the extraordinary environmental variation in the family.

Within the last thirty years numerous species of *Cladoniaceae* have experienced one or two changes in their current nomenclature, usually due to purely nomenclatural reasons. Many old names are still untypified and endanger the stability of the present nomenclature. In fact, I am aware of definite or potential nomenclatural changes in about 20 European species of *Cladonia*, if the *International code of botanical nomenclature* be strictly followed. I have published many necessary changes but refrained from the most unpleasant ones in the hope that the *Code* might be changed so that they could be avoided. Thus in *Cladoniaceae* the acceptance of the NCU proposals and subsequent approval of the present list really meet a great need.

This list thus includes several names which I know would definitely or at least probably have to be displaced by another name under the *Code*. It also includes type citations for all but half a dozen of the names, and in several cases new type designations.

The present list is a result of my studies on *Cladoniaceae* during the last 35 years, including visits to numerous herbaria in Eurasia, North and South America and Australasia. The outstanding monograph by E. A. Vainio (1887-1897) has been indispensable for locating the names and the original material. Other lichenologists working on *Cladoniaceae* (including its recent segregate *Cladiaceae* but not the *Heterodeaceae*) have also done much invaluable work on the taxonomy and nomenclature of this group. They include H. Sandstede, Y. Asahina, A. Evans, H. des Abbayes, R. Filson, T. Tønsberg, A. W. Archer, D. J. Galloway, and S. Stenroos.

Technical explanations

The list follows the general model of the lists included in this book, as explained in the general introduction. It may be mentioned that illustrations that are part of the protologue are not cited, except (rarely) for typification purposes.

Most of the type specimens cited have been examined by me. However, in several cases I have relied on other authors, especially with respect to duplicates (isotypes). I have listed many such duplicates, but completeness cannot be easily achieved (and is not necessary for the purpose of this list), especially in the case of widespread exsiccata (e.g., *Cladoniae exsiccatae* of H. Sandstede, distributed in 50 not fully identical sets).

The new neotype designations may call for some comments. They primarily concern names of taxa described by C. Linnaeus, W. Hudson, G. F. Hoffmann, H. G. Flörke and F. Eschweiler, whose herbaria were burnt or are otherwise incomplete and do not include the types in question. Some of the neotypified names could admittedly have been typified through references to J. J. Dillenius's pre-Linnaean drawings or other similar sources, but with uncertain results as to the stable application of the name in its traditional sense. In several cases I have therefore preferred to neotypify such names by specimens of more recent, widely distributed exsiccata, which have a known origin. Such specimens were selected from the country, and even province, where the original material came from, to secure the continued application of the name in its traditional meaning. However, the material is not in all cases from the immediate vicinity of the type locality, because it was felt more important that types should be present in well-known exsiccata sets and easily available for study in many herbaria and in different countries.

In very few cases (e.g. *Cladonia coniocraea*, *C. ochrochlora*, *C. macilenta*) the type designation definitely goes against the present rules, so as to keep the names in use in their traditional sense. Published type designations by other authors have always been accepted, however, as far as I am aware. Illustrations not backed by "typotypes" have been accepted as types only exceptionally (e.g., for *C. botrytes*). In the case of lichens the chemistry is often so important that a mere illustration cannot serve as a satisfactory type.

The type localities are given in terms of modern geographical and admininistrative units, whenever possible (some still requiring precision). The original geographical designations are added between quotation marks when they are much deviating (e.g., "Van Diemen's Land", now Tasmania), so that the type specimen can be recognized as such in the herbarium. The alternative option, to give the original type citation only, is often very misleading.

As to the selection of those names that are considered to be in current use, the list had naturally to take into account my taxonomic opinions, but also those of other workers as documented by their publications or submitted comments. I have therefore included many names which I am not presently accepting, and several names are listed at different ranks (e.g. *Cladonia elongata, C. gracilis* subsp. *elongata* and *C. gracilis* var. *elongata*) or in different combinations (e.g., both under *Cladonia* and *Cladina*). In one case, *Cladonia arbuscula* subsp. *beringiana* Ahti, a name is purposely omitted: Its exclusion will, once the list is approved, avoid its having to be adopted for a very common European taxon (*C. arbuscula* subsp. *squarrosa* (Wallroth) Ruoss) if the two subspecies will be united (as I expect). *C. borbonica* Nylander is also omitted, although I have adopted it in a recent paper (Ahti & Aptroot, 1992) as an older name for *C. poeciloclada* Abbayes; approval of the list would thus maintain the well-established name *C. poeciloclada* in use. Names of some so-called chemical taxa (based on characters of secondary chemistry only) are listed if recently used, often at varietal level. The classification and nomenclature of the subgenera and sections in *Cladonia* is not yet well established, but some familiar names are listed.

The standardized abbreviations designating herbaria have been expanded in several cases, to designate separately kept herbaria which are not listed as such, or not given abbreviations of their own, in the current edition of *Index herbariorum (herbaria)* (Holmgren & al., 1990). Such "unofficial" abbreviations are:

BERN-Frey (Herb. E. Frey in BERN);
BM-Ach (Herb. E. Acharius in BM);
FH-Dodge (Herb. C. W. Dodge in FH);
FH-Taylor (Herb. T. Taylor in FH);
FH-Tuck (Herb. E. Tuckerman in FH);
Herb. Kalb (Herb. K. Kalb, Lichenologisches Institut, Neumarkt/Oberpfalz, Germany);
Herb. Lai (Herb. M. J. Lai, Fu Jen University, Hsingchuang, Taiwan);
Herb. Huneck (Herb. S. Huneck, Halle, Germany);
LINN-Sm (Herb. J. E. Smith in LINN);
MW-Hoffm (Herb. G. F. Hoffmann in MW);
PC-Desm (Herb. J. B. H. J. Desmazières in PC);
PC-Fée (Herb. A. L. A. Fée in PC);
PC-Hue (Herb. A.-M. Hue in PC);
PC-Lenorm (Herb. S.-R. Lenormand in PC);
PC-Mont (Herb. J. P. F. C. Montagne in PC);
PC-Thur (Herb. G. A. Thuret in PC);
REN-Abbayes (Herb. H. des Abbayes, private property of Dr. J.-L. Massé, in REN);

S-Sw (Herb. O. Swartz in S);
UPS-Ach (Herb. E. Acharius in UPS);
UPS-Thunb (Herb. C. P. Thunberg in UPS).

Acknowledgements

I am much indebted to the following persons for corrections, additions and other help received during the preparation of this list: Dr. Alan W. Archer, Sydney; Dr. Irwin M. Brodo, Ottawa; Dr. Paula DePriest, Washington, DC; Dr. Jack A. Elix, Canberra; Mr. Aljos Farjon, Utrecht; Mr. Rex B. Filson, Booral, N.S.W.; Dr. David J. Galloway, London; Dr. Michael Giersberg, Rostock; Prof. Werner Greuter, Berlin; Mr. Samuel H. Hammer, Cambridge, MA; Prof. David L. Hawksworth, Egham, Surrey; Prof. Pekka Isoviita, Helsinki; Prof. Per Magnus Jörgensen, Bergen; Dr. Hiroyuki Kashiwadani, Tokyo; Mr. Jack R. Laundon, London; Dr. Engelbert Ruoss, Lucerne; Prof. Rolf Santesson, Uppsala; Mr. Thorsten Lumbsch, Essen; Dr. Harrie Sipman, Berlin; Dr. Soili Stenroos, Helsinki; Dr. Benito C. Tan, Cambridge, MA; and Prof. Jiang-chun Wei, Beijing.

Calathaspis I. M. Lamb & W. A. Weber

Caluthaspis devexa I. M. Lamb & W. A. Weber in Occas. Pap. Farlow Herb. Cryptog. Bot. 4: 1. Jul 1972. – Holotype: Papua New Guinea, Simbu, Mt. Wilhelm, Pindaunde Field Station, near Lake Aunde, 1967, *D. McVean 6782* (FH; isot.: CBG, COLO, LAE).

Cladia Nyl.

Cladia aggregata (Sw.) Nyl. in Compt. Rend. Hebd. Séances Acad. Sci. 83: 88. Jul 1876 [and in J. Linn. Soc., Bot., 15: 225. 1876]. – Basionym: *Lichen aggregatus* Sw., Prodr.: 147. 10 Jun - 29 Jul 1788. – Lectotype (Filson, 1981): Jamaica, *O. P. Swartz s.n.* (S; iso-lectot.: ?BM, ?G, ?H-ACH No. 1583, ?PC-Hue, ?UPS-Thunb).

Cladia corallaizon Filson in Victorian Naturalist 87: 324. 4 Nov 1970. – Holotype: Australia, New South Wales, Narrandera to West Wyalong road, 45 km N of Ardlethan, 1963, *R. Filson 5466* (MEL No. 25246).

Cladia ferdinandii (Müller Arg.) Filson in Victorian Naturalist 87: 325. 4 Nov 1970. – Basionym: *Cladonia ferdinandii* Müller Arg. in Flora 65: 293 (1 Jul 1882). – Holotype: Australia, Western Australia, near Esperance Bay, 1864-1890, *A. Dempster s.n.* (G).

Cladia fuliginosa Filson in Victorian Naturalist 87: 325. 4 Nov 1970. – Holotype: Australia, Tasmania, Cradle Mountain-Lake St. Clair National Park, Mt. Campbell, 1968, *R. Filson 10854* (MEL No. 1001725).

Cladia inflata (F. Wilson) D. J. Galloway in Nova Hedwigia 28: 476. 15 Apr 1977 ["1976"]. – Basionym: *Cladonia aggregata* var. *inflata* F. Wilson in Pap. & Proc. Roy. Soc. Tasmania 1892-1893: 153. 1893. – Lectotype (Galloway, 1977a): Australia, Tasmania, Maria Island, *R. A. Bastow s.n.* (NSW No. 2084).

Cladia moniliformis Kantvilas & Elix in Mycotaxon 29: 199. 31 Jul 1987. – Holotype: Australia, Tasmania, N of Sentinel Range, c. 4 km SE of Mt. Cullen, 1986, *G. Kantvilas & J. Jarman 169/86* (HO; isot.: BM, CBG, MEL).

Cladia retipora (Labill.) Nyl. in Compt. Rend. Hebd. Séances Acad. Sci. 83: 88. Jul 1876. – Basionym: *Baeomyces retiporus* Labill., Nov. Holl. Pl. 2: 110. Mar-Aug 1807 ["1804"]. – Lectotype (Filson, 1981): Australia, Tasmania ("Cap Van Diemen"), 1792, *J. J. H. Labillardière s.n.* (PC; isolectot.: BM, FH-Tuck, FI-W, G, LINN-Sm, PC).

Cladia schizopora (Nyl.) Nyl. in Rev. Bot. Bull. Mens. 6: 161. 1888. – Basionym: *Cladonia schizopora* Nyl., Syn. Meth. Lich. 1: 217. Apr 1860. – Holotype: Australia, Tasmania, *C. Stuart s.n.* (H-NYL No. 37591).

Cladia sullivanii (Müller Arg.) W. Martin in Trans. Roy. Soc. New Zealand, Bot. 2: 44. 30 Nov 1962. – Basionym: *Cladonia sullivanii* Müller Arg. in Flora 65: 294. 1 Jul

1882. – Holotype: Australia, Victoria, Grampian Mts., *D. Sullivan 10* (G; isot.: ?UPS).

Cladina Nyl.

Cladina sect. ***Impexae*** (Abbayes) Ahti in Beih. Nova Hedwigia 79: 30. 1984. – Basionym and type: see under *Cladonia* sect. *Impexae.*

Cladina sect. ***Tenues*** (Abbayes) Ahti in Beih. Nova Hedwigia 79: 38. 1984. – Basionym and type: see under *Cladonia* sect. *Tenues.*

Cladina aberrans (Abbayes) Hale & W. L. Culb. in Bryologist 73: 510. 17 Nov 1970. – Basionym and type: see under *Cladina stellaris* var. *aberrans.*

Cladina arbuscula (Wallr.) Hale & W. L. Culb. in Bryologist 73: 510. 17 Nov 1970. – Basionym and type: see under *Cladonia arbuscula.*

Cladina arbuscula subsp. ***squarrosa*** (Wallr.) Burgaz in Nova Hedwigia 55: 38. Aug 1992. – Basionym and type: see under *Cladonia arbuscula* subsp. *squarrosa.*

Cladina arcuata (Ahti) Ahti & Follmann in Philippia 4: 321. 15 Mar 1981. – Basionym and type: see under *Cladonia arcuata.*

Cladina argentea Ahti in Ann. Bot. Fenn. 23: 221. 26 Nov 1986. – Holotype: Venezuela, Bolívar, Distrito Piar, Macizo del Chimantá, Churítepui, 1985, *T. Ahti & al. 44914* (VEN; isot.: B, COL, H, MERF, NY, US).

Cladina azorica (Ahti) Ahti in Beih. Nova Hedwigia 79: 32. 1984. – Basionym and type: see under *Cladonia azorica.*

Cladina boliviana (Ahti) Ahti in Beih. Nova Hedwigia 79: 50. 1984. – Basionym and type: see under *Cladonia boliviana.*

Cladina ciliata (Stirt.) Trass in Folia Cryptog. Estonica 11: 6. 1 Mar 1979 ["1978"]. – Basionym and type: see under *Cladonia ciliata.*

Cladina ciliata var. *tenuis* (Flörke) Ahti & M. J. Lai in Ann. Bot. Fenn. 16: 234. 28 Nov 1979. – Basionym and type: see under *Cladonia ciliata* var. *tenuis.*

Cladina confusa (R. Sant.) Follmann & Ahti in Philippia 4: 321. 15 Mar 1981. – Basionym and type: see under *Cladonia confusa.*

Cladina conspicua Ahti in Ann. Bot. Fenn. 23: 224. 26 Nov 1986. – Holotype: Canada, Newfoundland, Humber East District, Topsail Uplands, 2.4 km N of Gaff Topsail, 1956, *T. Ahti 2997* (H; isot.: BM, CANL, M, MIN, NFLD, NMLU, TNS, UBC, UPS, US).

Cladina dendroides (Abbayes) Ahti in Beih. Nova Hedwigia 79: 38. 1984. – Basionym and type: see under *Cladonia dendroides.*

Cladina densissima Ahti in Beih. Nova Hedwigia 79: 38. 1984. – Holotype: Guyana, Pakaraima Mts., Mt. Aymatoi, 1981, *P. J. M. Maas & al. 5702* (U; isot.: H).

Cladina evansii (Abbayes) Hale & W. L. Culb. in Bryologist 73: 510. 17 Nov 1970. – Basionym and type: see under *Cladonia evansii.*

Cladina grisea (Ahti) Trass ex J. C. Wei in Acta Mycol. Sin. 5: 247. 1986. – Basionym and type: see under *Cladonia rangiferina* subsp. *grisea.*

Cladina halei Ahti in Ann. Bot. Fenn. 23: 224. 26 Nov 1986. – Holotype: Venezuela, Táchira, District Jáuregui, Mun. Vargas, Páramo de El Zumbador, 5 km S of El Cobre, 1985, *T. Ahti & M. López-Figueiras 43810* (H; isot.: B, BM, MERF, US, VEN).

Cladina imshaugii (Ahti) Ahti in Beih. Nova Hedwigia 79: 50. 1984. – Basionym and type: see under *Cladonia imshaugii.*

Cladina laevigata (Vain.) C. W. Dodge, Lich. Fl. Antarct.: 97. 1974 ["1973"]. – Basionym and type: see under *Cladonia laevigata.*

Cladina leiodea (H. Magn.) Ahti in Beih. Nova Hedwigia 79: 42. 1984. – Basionym and type: see under *Cladonia leiodea.*

Cladina luzonensis (Ahti) Gruezo in Kalikasan 6: 137. 1977. – Basionym and type: see under *Cladonia luzonensis.*

Cladina magnussonii (Ahti) Ahti in Beih. Nova Hedwigia 79: 44. 1984. – Basionym and type: see under *Cladonia magnussonii.*

Cladina mediterranea (P. A. Duvign. & Abbayes) Follmann & Hern.-Padr. in Philippia 3: 369. 30

Jun 1978. – Basionym and type: see under *Cladonia mediterranea*.

Cladina mitis (Sandst.) Hustich in Acta Geogr. (Helsinki) 12(1): 27. 1951. – Basionym and type: see under *Cladonia mitis*.

Cladina portentosa (Dufour) Follmann in Philippia 4: 34. 15 Mar 1979. – Basionym and type: see under *Cladonia portentosa*.

Cladina portentosa subsp. ***pacifica*** (Ahti) Ahti in Beih. Nova Hedwigia 79: 37. 1984. – Basionym and type: see under *Cladonia portentosa* subsp. *pacifica*.

Cladina pseudoëvansii (Asahina) Hale & W. L. Culb. in Bryologist 73: 510. 17 Nov 1970. – Basionym and type: see under *Cladonia pseudoëvansii*.

Cladina pycnoclada (Pers.) Leight. in Ann. Mag. Nat. Hist., ser. 3, 19: 122. 1867. – Basionym and type: see under *Cladonia pycnoclada*.

Cladina rangiferina (L.) Nyl. in Not. Sällsk. Fauna Fl. Fenn. Förh. 8: 110. Jun 1866. – Basionym and type: see under *Cladonia rangiferina*.

Cladina rangiferina subsp. ***abbayesii*** (Ahti) W. L. Culb. in Vězda, Lich. Sel. Exsicc. 76: 3 [No. 1883]. 1983. – Basionym and type: see under *Cladonia rangiferina* var. *abbayesii*.

Cladina rangiferina subsp. ***grisea*** (Ahti) Ahti & M. J. Lai in Ann. Bot. Fenn. 16: 234. 28 Nov 1979. – Basionym and type: see under *Cladonia rangiferina* subsp. *grisea*.

Cladina rotundata (Ahti) Ahti in Beih. Nova Hedwigia 79: 40. 1984. – Basionym and type: see under *Cladonia rotundata*.

Cladina sandstedei (Abbayes) Ahti in Beih. Nova Hedwigia 79: 40. 1984. – Basionym and type: see under *Cladonia sandstedei*.

Cladina skottsbergii (H. Magn.) Follmann in Philippia 4: 35. 15 Mar 1979. – Basionym and type: see under *Cladonia skottsbergii*.

Cladina sprucei (Ahti) Ahti in Beih. Nova Hedwigia 79: 40. 1984. – Basionym and type: see under *Cladonia sprucei*.

Cladina stellaris (Opiz) Brodo in Bryologist 79: 363. 15 Oct 1976. – Basionym and type: see under *Cladonia stellaris*.

Cladina stellaris var. ***aberrans*** (Abbayes) Ahti in Mycotaxon 28: 92. 14 Jan 1987. – Basionym: *Cladonia alpestris* f. *aberrans* Abbayes in Bull. Soc. Sci. Bretagne 16, Fasc. Hors Sér. 2: 93. Jul-Dec 1939. – Lectotype (Ahti, 1961): U.S.A., Alaska, St. Paul Island, 1897, *J. M. Macoun* in Macoun, Canad. Lich. No. 171 (W; isolectot.: CANL, US).

Cladina stygia (Fr.) Ahti in Beih. Nova Hedwigia 79: 45. 1984. – Basionym and type: see under *Cladonia stygia*.

Cladina submitis (A. Evans) Hale & W. L. Culb. in Bryologist 73: 510. 17 Nov 1970. – Basionym and type: see under *Cladonia submitis*.

Cladina subtenuis (Abbayes) Hale & W. L. Culb. in Bryologist 73: 510. 17 Nov 1970. – Basionym and type: see under *Cladonia subtenuis*.

Cladina tasmanica (Ahti) Ahti in Beih. Nova Hedwigia 79: 40. 1984. – Basionym and type: see under *Cladonia tasmanica*.

Cladina tenuis (Flörke) de Lesd. in Bull. Soc. Bot. France 54: 680. Feb 1908 ["1907"]. – Basionym and type: see under *Cladonia ciliata* var. *tenuis*.

Cladina terrae-novae (Ahti) Hale & W. L. Culb. in Bryologist 73: 511. 17 Nov 1970. – Basionym and type: see under *Cladonia terrae-novae*.

Cladina transindica (Ahti) Ahti in Beih. Nova Hedwigia 79: 44. 1984. – Basionym and type: see under *Cladonia transindica*.

Cladonia P. Browne

Cladonia subg. ***Cladina*** (Nyl.) Leight., Lich. Fl. Gr. Brit.: 66. Sep-Oct 1871. – Basionym: *Cladina* Nyl. in Not. Sällsk. Fauna Fl. Fenn. Förh. 8: 110. Jun 1866. – Lectotype (Ahti, 1961): *Cladina rangiferina* (L.) Nyl. (*Cladonia rangiferina* (L.) F. H. Wigg.).

Cladonia sect. ***Cocciferae*** (Delise) A. Evans in Trans. Connecticut Acad. Arts 30: 392. Jun 1930. – Basionym: *Cenomyce* [unranked] *Cocciferae* Delise in Duby, Bot. Gall.: 632. Mai 1830. – Holotype: *Cladonia coccifera* (L.) Willd.

Cladonia sect. ***Crustaceae*** Rabenh., Cladon. Eur.: 11. Nov 1860. – Lectotype (Ahti, 1984): *Cladonia rangiferina* (L.) F. H. Wigg.

Cladonia sect. ***Helopodium*** (Ach.) Stenroos in Ann. Bot. Fenn. 25: 118. 9 Aug 1988. – Basionym: *Baeomyces* [unranked] *Helopodium* Ach., Methodus: xxxvi, 326. Jan-Apr 1803. – Lectotype (designated here): *Baeomyces cariosus* (Ach.) Ach. (*Cladonia cariosa* (Ach.) Spreng.).

Cladonia sect. ***Impexae*** Abbayes in Bull. Soc. Sci. Bretagne 16, Fasc. Hors Sér. 2: 71. Jul-Dec 1939. – Holotype: *Cladonia impexa* Harm. [= *Cladonia portentosa* (Dufour) Coem.].

Cladonia sect. ***Perviae*** (Fr.) Mattick in Repert. Spec. Nov. Regni Vcg. 49: 160. 20 Dec 1940. – Basionym: *Cladonia* [unranked] *Perviae* Fr., Lichenogr. Eur. Reform.: 227. Jun-Jul 1831. – Lectotype (designated here): *Cladonia brachiata* (Fr.) Fr. [= *Cladonia cenotea* (Ach.) Schaer.].

Cladonia sect. ***Tenues*** Abbayes in Bull. Soc. Sci. Bretagne 16, Fasc. Hors Sér. 2: 71. Jul-Dec 1939. – Holotype: *Cladonia tenuis* (Flörke) Harm. [= *Cladonia ciliata* Stirt.].

Cladonia sect. ***Unciales*** (Delise) Oxner ex Ahti in Ann. Bot. Fenn. 23: 208. 26 Nov 1986. – Basionym: *Cenomyce* [unranked] *Unciales* Delise in Duby, Bot. Gall.: 620. Mai 1830. – Holotype: *Cenomyce uncialis* (L.) Ach. (*Cladonia uncialis* (L.) F. H. Wigg.).

Cladonia abbatiana Stenroos in Ann. Bot. Fenn. 28: 107. 3 Jul 1991. – Holotype: Madagascar, Antseranana, Res. Nat. Marojezy, 1972, *J.-L. Guillaumet 4084* (PC; isot.: H).

Cladonia abbreviatula G. Merr. in Bryologist 27: 21. 6 Mai 1924. – Holotype: U.S.A., Florida, Seminole Co., Sanford, 1922, *S. Rapp 572* (FH).

Cladonia aberrans (Abbayes) Stuckenb. in Bot. Mater. Otd. Sporov. Rast. Bot. Inst. Komarova Akad. Nauk S.S.S.R. 11: 12. 1956. – Basionym and type: see under *Cladina stellaris* var. *aberrans*.

Cladonia acuminata (Ach.) Norrl. in Norrlin & Nylander, Herb. Lich. Fenniae No. 57a [label]. ante Sep 1875 [and Index: 3. 1875; and in Flora 58: 447. 1 Oct 1875]. – Basionym: *Cenomyce pityrea* f. *acuminata* Ach., Syn. Lich. 254. 1814. – Lectotype (Stenroos & Ahti, 1991): Switzerland, *Schleicher 57-2* (H-ACH No. 1712E).

Cladonia acuta (Taylor) Hue in Nouv. Arch. Mus. Hist. Nat., ser. 3, 2: 253. 1890. – Basionym and type: see under *Cladonia rigida* var. *acuta*.

Cladonia adspersa Mont. & Bosch in Miquel, Pl. Jungh.: 456. 1855-1857. – Lectotype (designated here): Indonesia, Java, *F. W. Junghuhn* ex Herb. van den Bosch No. 94 (PC-Hue).

Cladonia ahtii Stenroos in Ann. Bot. Fenn. 26: 252. 3 Jul 1989. –

Holotype: Brazil, Rio Grande do Sul, Mun. Viamão, Morro do Coco, 1980, *H. S. Osorio 7794* (H; isot.: IFP, MVM).

Cladonia alaskana A. Evans in Bryologist 50: 35. 15 Apr 1947. – Holotype: U.S.A., Alaska, Cantwell, 1926, *L. J. Palmer 1438* (US).

Cladonia alinii Trass in Folia Cryptog. Estonica 11: 1. 1 Mar 1979 ["1978"]. – Holotype: Russia, Far East, Primor'e Terr., Sikhote-Alin Range, Mt. Chernyy, 1977, *H. Trass Pri-563* (TU; isot.: CANL, H, NY, US).

Cladonia alpina (Asahina) Yoshim. in J. Hattori Bot. Lab. 31: 198. 19 Nov 1968. – Basionym: *Cladonia floerkeana* var. *alpina* Asahina in J. Jap. Bot. 15: 665. Nov 1939. – Lectotype (Yoshimura, 1968): Japan, Hokkaido, Ishikari, Mt. Ashibetsu, 1935, *Y. Asahina 35726* (TNS).

Cladonia amaurocraea (Flörke) Schaer., Lich. Helv. Spic.: 34. 1823. – Basionym: *Capitularia amaurocraea* Flörke in Beitr. Naturk. 2: 334. 19 Sep 1810. – Lectotype (designated here): Austria, Salzburg, *H. G. Flörke 58* (BM).

Cladonia andesita Vain. in Hedwigia 38: (124). 26 Jun 1899. – Lectotype (Ahti & al. 1987): Colombia, near Bogotá, *J. Weir 39* (BM; isolectot.: TUR-V No. 18367).

Cladonia angustata Nyl. in Ann. Sci. Nat., Bot., ser. 4, 11: 236. Apr 1859. – Lectotype (Stenroos, 1986): Hawaii, 1851-1855,

J. Rémy s.n. (H-NYL No. 37978; isolectot.: FH-Tuck, PC).

Cladonia anitae W. L. Culb. & C. F. Culb. in Mycologia 74: 663. 21 Jul 1982. – Holotype: U.S.A., North Carolina, Onslow Co., between Folkstone and Holly Ridge, *W. L. Culberson & C. F. Culberson 18539* in Vězda, Lich. Sel. Exsicc. No. 1854 (DUKE; isot.: BM, CANL, DUKE, H, S, UPS, US).

Cladonia anserina Ahti in Ann. Bot. Fenn. 29: 68. 24 Apr 1992. – Holotype: Falkland Is., East Falklands, Goose Green, Fish Creek, 1963, *R. W. M. Corner 83* (AAS; isot.: H).

Cladonia apodocarpa Robbins in Rhodora 27: 211. 17 Feb 1926. – Lectotype (designated here): U.S.A., Massachusetts, Plymouth Co.: Wareham, 1924, *C. A. Robbins s.n.* (FH).

Cladonia arbuscula (Wallr.) Flot. in Wendt, Thermen Warmbrunn: 94. 1839. – Basionym: *Patellaria foliacea* var. *arbuscula* Wallr., Naturgesch. Säulchen-Flecht.: 169. 1829. – Lectotype (Ruoss & Ahti, 1985): Germany, Thuringia, Nordhausen, Wallroth's plate 261 (STR); photograph of part in Arnold, Lich. Exsicc. No. 1348 (G, H, M, STR, UPS).

Cladonia arbuscula subsp. *mitis* (Sandst.) Ruoss in Bot. Helv. 97: 260. 30 Dec 1987. – Basionym and type: see under *Cladonia mitis.*

Cladonia arbuscula subsp. ***squarrosa*** (Wallr.) Ruoss in Bot. Helv. 97: 260. 30 Dec 1987. – Basionym: *Patellaria coccinea* var. *squarrosa* Wallr., Naturgesch. Säulchen-Flecht.: 191. 1829. – Lectotype (Ruoss & Ahti, 1985): Germany, Harz Mts., Brocken, *F. W. Wallroth s.n.* (STR).

Cladonia arbuscula subsp. ***stictica*** Ruoss in Lichenologist 21: 37. Jan 1989. – Holotype: New Zealand, Otago, Mt. Maungatua, 1935, *J. S. Thomson 2146* (CHR; isot.: CHR, H, NMLU).

Cladonia arbuscula var. ***mitis*** (Sandst.) Sipman in Wetensch. Meded. Kon. Ned. Natuurhist. Ver. 124: 43. Jan 1978. – Basionym and type: see under *Cladonia mitis.*

Cladonia arcuata Ahti in Ann. Bot. Soc. Zool.-Bot. Fenn. "Vanamo" 32(1): 73. 25 Sep 1961. – Holotype: Bolivia, La Paz, Sur Yungas, between San Felipe and El Chaco, 1920, *E. Asplund 34* (S; isot.: UPS).

Cladonia artuata S. Hammer in Bryologist 96: 80. 22 Mar 1993. – Holotype: U.S.A., California, Mendocino Co., Jackson State Forest, 1986, *S. Hammer 1095* (FH; isot.: H, NY, SFSU, US, WIS).

Cladonia asahinae J. W. Thomson in J. Jap. Bot. 51: 361. Jan-Feb 1977 ["Dec 1976"]. – Holotype: U.S.A., Washington, Skagit Co., Mt. Erie on Fidalgo Island S of Anacortes, 1969, *J. W. Thomson 16296* (WIS).

Cladonia atlantica A. Evans in Trans. Connecticut Acad. Arts 35: 573. 1944. – Holotype: U.S.A., Connecticut, Middlesex Co., Middletown, 1932, *A. Evans 2966* (US).

Cladonia attendenda Vain. in Ann. Acad. Sci. Fenn., Ser. A, 15(6): 56. 1921. – Lectotype (designated here): The Philippines, Luzon, Benguet, Mt. Pulog, *H. M. Curran & al.* in Forestry Bur. No. 16261 (TUR-V No. 18378).

Cladonia aueri Räsänen in Ann. Bot. Soc. Zool.-Bot. Fenn. "Vanamo" 2(1): 53. 1932. – Lectotype (Ahti & Kashiwadani, 1984): Chile, Tierra del Fuego, Río Bueno, 1928, *H. Roivainen 1826* (H; isolectot.: BA, H, S).

Cladonia azorica Ahti in Ann. Bot. Soc. Zool.-Bot. Fenn. "Vanamo" 32(1): 36. 25 Sep 1961. – Holotype: Azores, Pico, *A. G. da Cunha & L. Sobrinho s.n.* (LISU).

Cladonia bacillaris Genth, Fl. Nassau: 406. Sep-Oct 1835. – Neotype (designated here): England, Durham, Cleveland, Ayton Moor, *W. Mudd* in Mudd, Monogr. Brit. Cladon. No. 70 (BM; isoneot.: FH, UPS).

Cladonia bacilliformis (Nyl.) Glück in Verh. Naturhist.-Med. Vereins Heidelberg, ser. 2, 6: 97. 1899. – Basionym: *Cladonia carneola* var. *bacilliformis* Nyl., Syn. Meth. Lich. 1: 201. Apr 1860. – Holotype: Finland, Uusimaa, Helsinki (Helsingfors), 1858, *W. Nylander s.n.* (H; isot.: BM).

Cladonia bangii Ahti in Ann. Bot. Fenn. 23: 208. 26 Nov 1986. – Holotype: Bolivia, near Tolapampa, 1901, *R. S. Williams 2259* (H; isot.: NY).

Cladonia beaumontii (Tuck.) Vain. in Acta Soc. Fauna Fl. Fenn. 10: 455. Dec 1894. – Basionym: *Cladonia santensis* var. *beaumontii* Tuck., Syn. N. Amer. Lich. 1: 245. Jan-Feb 1882. – Holotype: U.S.A., Alabama "septentrionalis" (probably Macon Co., Tuskegee), 1853, *J. F. Beaumont 25* (FH-Tuck).

Cladonia bellidiflora (Ach.) Schaer., Lich. Helv. Spic.: 21. 1823. – Basionym: *Lichen bellidiflorus* Ach., Lichenogr. Suec. Prodr.: 194. 1799 ["1798"]. – Lectotype (designated here): Sweden, *E. Acharius s.n.* (H-ACH No. 1569A).

Cladonia bellidiflora var. *crassa* Räsänen in Ann. Bot. Soc. Zool.-Bot. Fenn. "Vanamo" 20(3): 20. 1944. – Lectotype (Ahti & Kashiwadani, 1984): Chile, Aisén, Lago San Martín Glaciares, 1933, *A. Donat 13* (H; isolectot.: H).

Cladonia berghsonii Asperges in Cryptog. Bryol. Lichénol. 2: 349. 20 Sep 1981. – Holotype: Belgium, Hautes-Ardennes, Malmédy-Bévercé, 1979, *M. Asperges 2951* (BR; isot.: H, U).

Cladonia bimberiensis A. W. Archer in Muelleria 6: 93. 29 Mai 1985. – Holotype: Australia, Australian Capital Territory, Mt. Bimberi, 49 km SW of Canberra,

1979, *H. Streimann 9743* (CBG; isot.: H, TNS, US).

Cladonia boivinii Vain. in Acta Soc. Fauna Fl. Fenn. 4: 408. 3-31 Dec 1887. – Holotype: Comores, *L. H. Boivin s.n.* (PC; isot.: FH-Dodge, TUR-V No. 16872).

Cladonia boliviana Ahti in Ann. Bot. Soc. Zool.-Bot. Fenn. "Vanamo" 32(1): 131. 25 Sep 1961. – Holotype: Bolivia, La Paz, 1889, *M. Bang 20* (UPS; isot.: BM, CANL [as *20a*], NY, REN-Abbayes, US, WRSL).

Cladonia borealis Stenroos in Ann. Bot. Fenn. 26: 160. 3 Jul 1989. – Holotype: Finland, Etelä-Häme, Ylöjärvi, Pengonpohja, 1905, *A. A. Sola s.n.* (H; isot.: BM).

Cladonia boryi Tuck. in Proc. Amer. Acad. Arts 1: 246. Mar-Apr 1848. – Lectotype (Ahti, 1983): U.S.A., Massachusetts, Plymouth Co., Hingham, 1839, *J. L. Russell s.n.* (FH-Tuck).

Cladonia botryocarpa G. Merr. ex Sandst., Cladon. Exsicc., Uebers.: 52. Jul 1930. – Lectotype (designated here): U.S.A., Florida, Seminole Co., Sanford, 1923, *S. Rapp* in Sandstede, Cladon. Exsicc. No. 1142 (TUR-V No. 15148; isolectot.: BM, FH, TUR-V No. 15150, UPS, US).

Cladonia botrytes (K. G. Hagen) Willd., Fl. Berol. Prodr.: 365. 1787. – Basionym: *Lichen botrytes* K. G. Hagen, Tent. Hist. Lich.: 121, t. 2, f. 9. 1782. – Lectotype (designated here): Russia, Kaliningrad Region (East Prus-

sia), Wilke ("in silua Wilky") and near Tiefensee ("Tieffensee"), *K. G. Hagen* in Hagen, Tent. Hist. Lich.: t. 2, fig. 9. 1782.

Cladonia brasiliensis (Nyl.) Vain. in Acta Soc. Fauna Fl. Fenn. 10: 418. Dec 1894. – Basionym: *Cladonia substraminea* var. *brasiliensis* Nyl. in Flora 52: 117. 30 Mar 1869. – Lectotype (Stenroos, 1989b): Brazil, "Rio de Janeiro" (probably Amazonas), "1867, *A. Glaziou 1171* or *1878*" [probably 1849-1855, *R. Spruce*] (M; isolectot.: BM, G, H-NYL No. 38551, PC-Fée).

Cladonia brevis (Sandst.) Sandst., Cladon. Exsicc. No. 401. 1 Jul 1919. – Basionym: *Cladonia verticillata* [unranked] *cervicornis* f. *brevis* Sandst., Cladon. Exsicc. No. 234. Aug 1918. – Lectotype (designated here): Germany, Niedersachsen, Oldenburg, Sandhatten, 1918, *H. Sandstede* in Sandstede, Cladon. Exsicc. No. 234 (H; isolectot.: B, BM, FH, UPS).

Cladonia buckii R. C. Harris, Florida Lich.: 5. 21 Dec 1990. – Holotype: U.S.A., Florida, Franklin Co., Apalachicola National Forest, c. 8 mi. S of Sumatra, 1990, *W. R. Buck 17871* (NY).

Cladonia caespiticia (Pers.) Flörke, De Cladon.: 8. 1827. – Basionym: *Baeomyces caespiticius* Pers. in Ann. Bot. (Usteri) 7: 155. 1794. – Lectotype (designated here): The Netherlands?, *C. H. Persoon s.n.* (H-ACH No. 1722A).

Cladonia calycantha Nyl. in Ann. Sci. Nat., Bot., ser. 4, 11: 209. Apr

1859. – Lectotype (Ahti, 1983): Peru, 1839-1840, *J. Gay s.n.* (H-NYL No. 38557; isolectot.: G, M, PC).

Cladonia calyciformis Nuno in J. Jap. Bot. 47: 161. Jun 1972. – Holotype: Thailand, Phu Kradung, 1964, *S. Kurokawa 1855* (TNS).

Cladonia candelabrum (Bory) Nyl. in Mém. Soc. Sci. Nat. Cherbourg 5: 95. 24-30 Mar 1858 ["1857"]. – Basionym: *Lichen candelabrum* Bory, Voy. Iles Afrique 3: 103. Sep 1804. – Holotype: Réunion ("Insula Borbonia"), Plaine-des-Chicots, *J. Bory s.n.* (PC-Thur; isot.: ?H-ACH No. 1639B, ?TUR-V No. 13623).

Cladonia capitellata (Hook. f. & Taylor) C. Bab. in Hooker, Fl. Nov.-Zel. 2: 296. 9 Feb 1855. – Basionym: *Cenomyce capitellata* Hook. f. & Taylor in London J. Bot. 3: 652. Dec 1844. – Lectotype (Galloway, 1985): New Zealand, *J. D. Hooker s.n.* (BM; isolectot.: BM, FH-Taylor, FH-Tuck).

Cladonia capitellata var. ***interhiascens*** (Nyl.) Sandst. in Repert. Spec. Nov. Regni Veg., Beih., 103: 36. Dec 1938. – Basionym: *Cladina interhiascens* Nyl. in Compt. Rend. Hebd. Séances Acad. Sci. 83: 87. Jul 1876. – Lectotype (Archer, 1986b): New Zealand, Campbell Island, *M. Filhol s.n.* (H-NYL No. 37608; isolectot.: PC, TUR-V No. 13620 & 13621).

Cladonia capitellata var. ***squamatica*** A. W. Archer in Proc. Linn. Soc. New South Wales 108: 193. 21 Mai 1986. – Holotype: Australia, New South Wales, Tinderry Range, 1981, *A. W. Archer s.n.* (NSW; isot.: MEL).

Cladonia carassensis Vain. in Acta Soc. Fauna Fl. Fenn. 4: 313. 3-31 Dec 1887. – Lectotype (designated here): Brazil, Minas Gerais, Caraça, 1885, *E. A. Vainio s.n.* (TUR-V No. 15167; isolectot.: FH, G, REN-Abbayes, US).

Cladonia carassensis subsp. ***japonica*** (Vain.) Asahina in J. Jap. Bot. 34: 348. Nov 1959. – Basionym and type: see under *Cladonia japonica*.

Cladonia caribaea Stenroos in Ann. Bot. Fenn. 26: 256. 31 Aug 1989. – Holotype: Guatemala, Baja Verapaz, c. 15 km S of Purulhá, 1979, *K. Kalb & G. Plöbst s.n.* (H; isot.: Herb. Kalb).

Cladonia cariosa (Ach.) Spreng., Syst. Veg. 4(1): 272. 1-7 Jun 1827. – Basionym: *Lichen cariosus* Ach., Lichenogr. Suec. Prodr.: 198. 1799 ["1798"]. – Lectotype (Stenroos & al., 1992): Sweden, *E. Acharius s.n.* (H-ACH No. 1577A; isolectot.: ?BM, ?UPS-Ach).

Cladonia carneola (Fr.) Fr., Lichenogr. Eur. Reform.: 233. Jun-Jul 1831. – Basionym: *Cenomyce carneola* Fr., Sched. Crit. Lich. Suec. 1-4: 23. 7 Mai 1825. – Lectotype (Stenroos & al., 1992): Sweden, *E. M. Fries* in Fries, Lich. Suec. Exsicc. No. 115 (UPS; isolectot.: BM, FH-Tuck).

Cladonia caroliniana Tuck. in Amer. J. Sci. Arts, ser. 2, 25: 427. Mai 1858. – Lectotype (Ahti, 1973): U.S.A., North Carolina, Forsyth Co., Salem, (Bennett's Rocks?), *L. von Schweinitz s.n.* (FH-Tuck; isolectot.: UPS).

Cladonia cartilaginea Müller Arg. in Flora 63: 260. 11 Jun 1880. – Holotype: Venezuela, near Caracas, *A. Ernst 3* (G; isot.: G).

Cladonia celata A. W. Archer in Muelleria 5: 271. 5 Apr 1984. – Holotype: Australia, New South Wales, Tinderry Range, Captain's Flat Road, 10 km E of Michelago, 1981, *A. W. Archer 1185* (MEL; isot.: H).

Cladonia cenotea (Ach.) Schaer., Lich. Helv. Spic.: 35. 1823. – Basionym: *Baeomyces cenoteus* Ach., Methodus: 345. Jan-Apr 1803. – Lectotype (designated here): Sweden, *E. Acharius s.n.* (H-ACH No. 1572C).

Cladonia centrophora Müller Arg. in Flora 70: 286. 21 Jul 1887. – Holotype: South Africa, Cape Province, Cape of Good Hope, Table Mountain, *F. Wilms 112* (G).

Cladonia ceratophylla (Sw.) Spreng., Syst. Veg. 4(1): 271. 1-7 Jun 1827. – Basionym: *Lichen ceratophyllus* Sw., Prodr.: 147. 10 Jun - 25 Jul 1788. – Lectotype (designated here): Jamaica, *O. P. Swartz s.n.* (S-Sw; isolectot.: BM, G, H-ACH No. 1576, S, UPS-Ach).

Cladonia ceratophyllina (Nyl.) Vain. in Nouv. Arch. Mus. Hist. Nat., ser. 3, 10: 273. 1898. – Basionym: *Cladonia degenerans* var. *ceratophyllina* Nyl. in Ann. Sci. Nat., Bot., ser. 4, 11: 249. Apr 1859. – Lectotype (Ahti & Aptroot, 1992): Réunion, *P. Lepervanche-Mézières 54* (H-NYL No. 38676; isolectot.: G, PC-Thur).

Cladonia cervicornis (Ach.) Flot. in Jahresber. Schles. Ges. Vaterl. Cult. 27: 105. 1849. – Basionym: *Lichen cervicornis* Ach., Lichenogr. Suec. Prodr.: 184. 1799 ["1798"]. – Lectotype (designated here): Sweden, *E. Acharius s.n.* (H-ACH No. 1672B-C; isolectot.: BM-Ach No. 722A)

Cladonia cervicornis subsp. ***mawsonii*** (C. W. Dodge) Stenroos & Ahti in Ann. Bot. Fenn. 27: 320. 8 Jan 1991. – Basionym and type: see under *Cladonia mawsonii.*

Cladonia cervicornis subsp. ***pulvinata*** (Sandst.) Ahti in Ann. Bot. Fenn. 20: 5. 28 Apr 1983. – Basionym: *Cladonia verticillata* f. *pulvinata* Sandst., Cladon. Exsicc. No. 233. 15 Aug 1918. – Lectotype (Ahti, 1983): Germany, Niedersachsen, Oldenburg, Markhausen, 1916, *H. Sandstede* in Sandstede, Cladon. Exsicc. No. 233 (H; isolectot.: BM, FH, UPS).

Cladonia cervicornis subsp. ***verticillata*** (Hoffm.) Ahti in Lichenologist 12: 126. Feb 1980. – Basionym and type: see under *Cladonia verticillata.*

Cladonia cervicornis var. ***verticillata*** (Hoffm.) Flot. in Jahresber. Schles. Ges. Vaterl. Cult. 27: 105.

1849. – Basionym and type: see under *Cladonia verticillata*.

Cladonia chlorophaea (Flörke ex Sommerf.) Spreng., Syst. Veg. 4(1): 273. 1-7 Jun 1827. – Basionym: *Cenomyce chlorophaea* Flörke ex Sommerf., Suppl. Fl. Lapp.: 130. 1826. – Lectotype (designated here): Norway, Nordland, Saltdalen, *C. Sommerfelt s.n.* (O).

Cladonia chondrotypa Vain. in Acta Soc. Fauna Fl. Fenn. 4: 449. 3-31 Dec 1887. – Lectotype (designated here): Brazil, Minas Gerais, Caraça, 1885, *E. A. Vainio s.n.* (TUR-V No. 15186; isolectot.: FH, G, TUR-V, US).

Cladonia ciliata Stirt. in Scott. Naturalist (Perth) 3: 308. 1888. – Holotype: Scotland, Kirkcudbright, New Galloway, Knocknalling Wood, 1884, *J. McAndrew 64* (GLAM; isot.: BM).

Cladonia ciliata var. ***tenuis*** (Flörke) Ahti in Biblioth. Lichenol. 9: 48. 1977. – Basionym: *Cladonia rangiferina* var. *tenuis* Flörke, De Cladon.: 164. Pasq. 1828. – Lectotype (Ahti, 1961): Germany (?), Herb. H. G. Flörke No. 108 (H).

Cladonia cinereorubens Abbayes in Rev. Bryol. Lichénol. 16: 87. 1947. – Holotype: Madagascar, Ambatofuiandrahana, Itremo, 1938, *R. Decary s.n.* (PC; isot.: REN-Abbayes).

Cladonia coccifera (L.) Willd., Fl. Berol. Prodr.: 361. 1787. – Basionym: *Lichen cocciferus* L., Sp. Pl.: 1151. 1 Mai 1753. – Lectotype (designated here): [Sweden?], Herb. Linnaeus No. 1273.215 (LINN).

Cladonia colombiana Sipman in Proc. Kon. Ned. Akad. Wetensch., C, 82: 236. 1979. – Holotype: Colombia, Boyacá, Páramo de la Sarna between Sogamoso and Vado Hondo, 1973, *A. M. Cleef 9382* (COL; isot.: H, U, UPS, US).

Cladonia confusa R. Sant. in Ark. Bot. 30A(10): 13. 15 Dec 1942. – Holotype: Ecuador, Imbabura, Lake Cuicocha, Islote Chica, 1939, *E. Asplund L107* in R. Santesson, Lich. Austroamer. Herb. Regnell. No. 351 (S; isot.: BM, G, H, S, UPS).

Cladonia congesta Ahti in Ann. Bot. Fenn. 23: 209. 26 Nov 1986. – Holotype: Vietnam, Cha Pa, 1930, *W. P. Lowe s.n.* (BM).

Cladonia coniocraea (Flörke) Spreng., Syst. Veg. 4(1): 272. 1-7 Jun 1827. – Basionym: *Cenomyce coniocraea* Flörke, Deutsche Lich. 7: 14. 1821. – Neotype (designated here): Sweden, Närke, Svennevad, Korsmon, 1950, *G. Kjellmert* in Magnusson, Lich. Sel. Scand. Exsicc. No. 388 (UPS; isoneot.: B, H).

Cladonia conista A. Evans in Trans. Connecticut Acad. Arts 30: 472. Jun 1930. – Lectotype (designated here): U.S.A., Connecticut, New Haven Co., North Branford, 1927, *A. Evans & F. A. Musch 1269* (US).

Cladonia connexa Vain. in Acta Soc. Fauna Fl. Fenn. 4: 288. 3-31 Dec 1887. – Lectotype (designated here): Brazil, Minas Gerais, Caraça, 1885, *E. A. Vainio s.n.* (TUR-V No. 15012).

Cladonia consimilis Vain. in Acta Soc. Fauna Fl. Fenn. 4: 303. 3-31 Dec 1887. – Lectotype (designated here): Brazil, Minas Gerais, Caraça, 1885, *E. A. Vainio s.n.* (TUR-V No. 15032).

Cladonia convoluta (Lam.) Anders, Strauch- & Blattflechten Nordböhm.: 29. 1906 [and in Mitth. Nordböhm. Excursions-Clubs 29: 144. 1906]. – Basionym: *Lichen convolutus* Lam., Encycl. 3: 500. 13 Feb 1792. – Lectotype (designated here): France, Paris, Bois de Boulogne, *J. Deslongchamps s.n.* (P-LA).

Cladonia corallifera (Kunze) Nyl. in Flora 57: 70. 11 Feb 1874. – Basionym: *Cenomyce corallifera* Kunze in Weigelt, exsicc. [unnumbered printed label]. 1827-1828. – Lectotype (designated here): Suriname, 1827, *Weigelt s.n.* (UPS; isolectot.: FH, G, LE, MB, TNS, TUR-V No. 14165, WRSL).

Cladonia corniculata Ahti & Kashiw. in Inoue, Stud. Cryptog. S. Chile: 136. 30 Aug 1984. – Holotype: Chile, Llanquihue, Cerro Pavilo, 16 km W of Tegualda, 1981, *H. Kashiwadani 17882* (TNS; isot.: H, SGO).

Cladonia cornuta (L.) Hoffm., Descr. Pl. Cl. Crypt.: t. 25(1). 1794. – Basionym: *Lichen cor-nutus* L., Sp. Pl.: 1152. 1 Mai 1753. – Lectotype (designated here): [Sweden?], Herb. Linnaeus No. 1273.223, lower specimen (LINN).

Cladonia cornuta subsp. *groenlandica* (Å. E. Dahl) Ahti in Ann. Bot. Fenn. 17: 221. 4 Sep 1980. – Basionym and type: see under *Cladonia cornuta* var. *groenlandica*.

Cladonia cornuta var. *groenlandica* Å. E. Dahl in Meddel. Grønland 150(2): 100. 4 Feb 1950. – Holotype: Greenland, Julianehaab District, Agdluitsoq (Lichtenaufjord), Sletten, 1937, Å. *E. Dahl s.n.* (O; isot.: O).

Cladonia corymbescens Nyl. ex Leight. in Ann. Mag. Nat. Hist., ser. 3, 18: 407. Nov 1866. – Lectotype (Ahti in Stenroos, 1988): New Caledonia, Mont de M'bée, 1855-1860, *E. Vieillard 1785* (PC; isolectot.: G, H-NYL No. 38414).

Cladonia corymbites Nyl. in J. Bot. 15: 225. Aug 1877. – Lectotype (designated here): Costa Rica, Cartago, near Angostura, 1875, *H. Polakowsky 472* (H-NYL No. 39577; isot.: FH-Dodge, H-NYL No. 39578, US, WRSL).

Cladonia corymbosula Nyl. in Flora 59: 560. 11 Dec 1876. – Holotype: Cuba, N. Sophia, 1856-1858, *C. Wright* in Lich. Ins. Cubae, ser. 2, No. 93 (H-NYL No. 38446; isolectot.: FH-Tuck, G, TUR-V No. 17291).

Cladonia crassiuscula Ahti in Ann. Bot. Fenn. 23: 210. 26 Nov 1986. – Holotype: Brazil, Pará, Serra do Cachimbo, Base Aérea do Cachimbo, 1983, *L. Brako & M. J. Dibben 5853* (INPA; isot.: B, H, NY, TNS, US, VEN).

Cladonia crispata (Ach.) Flot. in Wendt, Thermen Warmbrunn: 93. 1839. – Basionym: *Baeomyces turbinatus* var. *crispatus* Ach., Methodus: 341. Jan-Apr 1803. – Lectotype (Stenroos, 1988): Sweden (H-ACH No. 1641B).

Cladonia crispata var. *cetrariiformis* (Delise) Vain. in Acta Soc. Fauna Fl. Fenn. 4: 392. 3-31 Dec 1887. – Basionym: *Cenomyce gracilis* var. *cetrariiformis* Delise in Duby, Bot. Gall.: 625. Mai 1830. – Lectotype (designated here): France, Calvados, Vire, 1824, *D. Delise s.n.* (PC-Thur).

Cladonia crispatula (Nyl.) Ahti in Lichenologist 9: 14. Apr 1977. – Basionym: *Cladina rangiferina* var. *crispatula* Nyl. in Flora 52: 117. 30 Mar 1869. – Lectotype (Ahti, 1977): Brazil, Rio de Janeiro, Serra dos Orgãos, 1867, *A. Glaziou 1869* (H-NYL No. 37627; isolectot.: BM, H, M, PC, SP, UPS).

Cladonia cristatella Tuck. in Amer. J. Sci. Arts, ser. 2, 25: 428. Mai 1858. – Lectotype (designated here): U.S.A., New Hampshire, White Mts., 1849, *W. Oakes s.n.* (FH-Tuck).

Cladonia cryptochlorophaea Asahina in J. Jap. Bot. 16: 711. 15 Dec 1940. – Lectotype (designated here): Germany, Niedersachsen, Oldenburg, Kaihausermoor, 1917, *H. Sandstede* in Sandstede, Cladon. Exsicc. No. 237 (TNS; isolectot.: FH, H, UPS).

Cladonia cyanipes (Sommerf.) Nyl. in Mém. Soc. Sci. Nat. Cherbourg 5: 95. 24-30 Mar 1858 ["1857"]. – Basionym: *Cenomyce cyanipes* Sommerf. in Kongel. Norske Videnskabersselsk. Skr. 19de Aarhundr. 2: 62. 1826 [& Suppl. Fl. Lapp.: 129. 1826]. – Lectotype (designated here): Norway, Nordland, Saltdalen, *C. Sommerfelt s.n.* (O; isolectot.: UPS).

Cladonia cyathomorpha Stirt. ex Walt. Watson in J. Bot. 73: 156. Jun 1935. – Lectotype (Jølle, 1977): Scotland, Kirkcudbright, New Galloway, near Loch Dungeon, 1881, *J. McAndrew s.n.* (E; isolectot.: BM, GLAM).

Cladonia cylindrica (A. Evans) A. Evans in Rhodora 52: 116. 10 Mai 1950. – Basionym: *Cladonia borbonica* f. *cylindrica* A. Evans in Trans. Connecticut Acad. Arts 30: 480. Jun 1930. – Lectotype (designated here): U.S.A., Connecticut, Middlesex Co., Chester, 1928, *A. Evans & F. A. Musch 2228* (US).

Cladonia dactylota Tuck. in Amer. J. Sci. Arts, ser. 2, 28: 201. 1859. – Lectotype (designated here): Cuba, Niñañina, 1857, *C. Wright* in Wright, Lich. Ins. Cubae No. 30 (FH-Tuck; isolectot.: BM, G, H-NYL No. 38544, UPS).

Cladonia dahliana Kristinsson in Lichenologist 6: 141. Oct 1974. – Holotype: Iceland, Thjórsárver, Hestalda, 1973, *H. Kristinsson 24539* (ICEL; isot.: AMNH, DUKE).

Cladonia decorticata (Flörke) Spreng., Syst. Veg. 4(1): 271. 1-7 Jun 1827. – Basionym: *Capitularia decorticata* Flörke in Beitr. Naturk. 2: 297. 19 Sep 1810. – Neotype (designated here): Germany, Berlin, *H. G. Flörke* in Flörke, Deutsche Lich. No. 75 (UPS; isoneot.: BM).

Cladonia deformis (L.) Hoffm., Deutschl. Fl. 2: 120. 1796. – Basionym: *Lichen deformis* L., Sp. Pl.: 1152. 1 Mai 1753. – Neotype (designated here): Sweden, Uppland, Värmdön, Hasseludden, 1915, *G. O. A. Malme* in Malme, Lich. Suec. Exsicc. No. 533 (S; isoneot.: B, H, S).

Cladonia dehiscens Vain. in Nouv. Arch. Mus. Hist. Nat., ser. 3, 10: 271. 1898. – Holotype: Vietnam, "Tonkin", *R.-P. Bon 6236* (TUR-V No. 17287; isot.: PC-Hue).

Cladonia delavayi Abbayes in Candollea 16: 203. Jul 1958. – Holotype: China, Yunnan, Mt. Tsong Chou, above Ca-ti, 1885, *R. P. Delavay 1558* (PC).

Cladonia dendroides (Abbayes) Ahti in Ann. Bot. Soc. Zool.-Bot. Fenn. "Vanamo" 32(1): 29. 25 Sep 1961. – Basionym: *Cladonia sandstedei* f. *dendroides* Abbayes in Bull. Soc. Sci. Bretagne 16, Fasc. Hors Sér. 2: 99. Jul-Dec 1939. – Lectotype (Ahti, 1961):

Brazil, Paraná, Jacareí ("Jacarehy"), 1914, *P. Dusén 15231* in Malme, Lich. Austroamer. Herb. Regnell. No. 226 (PC; isot.: G, H, R, S, UPS, US).

Cladonia didyma (Fée) Vain. in Acta Soc. Fauna Fl. Fenn. 4: 137. 3-31 Dec 1887. – Basionym: *Scyphophorus didymus* Fée, Essai Crypt. Ecorc.: cxviii, ci. 29 Jan 1825. – Lectotype (designated here): Dominican Republic ("Santo Domingo"), *D. Poiteau s.n.* (G; isolectot.: G, PC-Mont, UPS).

Cladonia didyma var. ***vulcanica*** (Zoll. & Moritzi) Vain. in Acta Soc. Fauna Fl. Fenn. 4: 145. 3-31 Dec 1887. – Basionym and type: see under *Cladonia vulcanica*.

Cladonia digitata (L.) Hoffm., Deutschl. Fl. 2: 124. 1796. – Basionym: *Lichen digitatus* L., Sp. Pl.: 1152. 1 Mai 1753. – Neotype (designated here): Sweden, Östergötland (Ostrogothia), *C. Stenhammar* in Stenhammar, Lich. Suec. Exsicc., ed. 2, No. 195 (UPS; isoneot.: BM, H).

Cladonia dilleniana Flörke, De Cladon.: 138. Pasq. 1828. – Holotype: Turks and Caicos Islands, Island Providenciales, 1725-1726, *M. Catesby* in Dillenius, Hist. Musc.: t. 16, f. 23. 1742.

Cladonia dimorpha S. Hammer in Mycotaxon 37: 339. 23 Apr 1990. – Holotype: U.S.A., California, Humboldt Co., Honeydew Road near Rockefeller Memorial Redwood Grove, 1988, *S. Hammer 1130* (FH; isot.: SFSU).

Cladonia dimorphoclada Robbins in Sandstede, Cladon. Exsicc. No. 1882. 20 Mar 1929. – Lectotype (Ahti, 1973): U.S.A., North Carolina, New Hanover Co., Wrightsville, 1928, *A. Evans 218* in Sandstede, Cladon. Exsicc. No. 1882 (H; isolectot.: BM, FH, TUR-V, UPS, US).

Cladonia diplotypa Nyl. in Flora 45: 475. 16 Oct 1862. – Lectotype (Ahti, 1977): Cameroon, Mt. Cameroon, 1800 m, *G. Mann 6* (H-NYL No. 38411; isolectot.: BM, H-NYL, US).

Cladonia dissimilis (Asahina) Asahina, Lich. Japan 1: 190. 1950. – Basionym: *Cladonia verticillata* subsp. *dissimilis* Asahina in J. Jap. Bot. 16: 466. 1940. – Holotype: Japan, Prov. Ettyu, Mt. Tateyama, 1923, *Y. Asahina s.n.* (TNS; isot.: US).

Cladonia divaricata Nyl., Syn. Meth. Lich. 1: 214. Apr 1860. – Lectotype (designated here): Brazil (PC; isolectot.: G, H-NYL No. 37600, PC, TUR-V No. 13624).

Cladonia diversa Asperges, Cladon. Sect. Coccif. Belg. 2: 364. 1983. – Holotype: Belgium, Kempisch district, Kalmthout, Van Ganzenven, 1974, *M. Asperges 2498* (BR; isot.: BM, H, U).

Cladonia dubia Abbayes in Bull. Inst. Franç. Afrique Noire 14: 22. 1952. – Lectotype (designated here): Ivory Coast, Man, Mont Tonkoui, Rocher aux Sacrifices, 1948, *H. Abbayes s.n.* (REN-Abbayes; isolectot.: UPS).

Cladonia ecmocyna Leight. in Ann. Mag. Nat. Hist., ser. 3, 18: 406. Nov 1866. – Lectotype (Ahti, 1980): Russia, Murmansk Region, Kola Peninsula, Svyatoy nos, 1863, *N. I. Fellman* in Fellman, Lich. Arct. No. 28 (BM; isolectot.: H).

Cladonia ecmocyna subsp. *intermedia* (Robbins) Ahti in Ann. Bot. Fenn. 17: 227. 4 Sep 1980. – Basionym: *Cladonia elongata* f. *intermedia* Robbins in Rhodora 33: 137. 1 Jun 1931. – Holotype: U.S.A., Wyoming, Yellowstone National Park, Canyon Junction, 1927, *S. F. Blake s.n.* (FH; isot.: US).

Cladonia elongata (Wulfen) Hoffm., Deutschl. Fl. 2: 419. 1796. – Basionym and type: see under *Cladonia gracilis* subsp. *elongata.*

Cladonia enantia Nyl., Rev. Cladon. Zwackh: 1. Mar 1888 [and Lich. Nov. Zel.: 18. Sep 1888]. – Lectotype (Galloway, 1985): New Zealand, 1867, *C. Knight 204* (H-NYL No. 38740; isolectot.: BM, H-NYL, TUR-V, ZT).

Cladonia evansii Abbayes in J. Bot. 76: 351. Dec 1938 [and in Bull. Soc. Sci. Bretagne 16, Fasc. Hors Sér. 2: 71. Jul-Dec 1939]. – Lectotype (designated here): Cuba, *C. Wright* in Wright, Lich. Ins. Cubae No. 39 (PC-Hue; isolectot.: FH, FH-Tuck, G, M, UPS, US).

Cladonia farinacea (Vain.) A. Evans in Rhodora 52: 95. 10 Mai 1950. – Basionym: *Cladonia furcata* var. *farinacea* Vain. in J. Bot.

(Morot) 1: 283. 1 Nov 1887 [and *Cladonia furcata* f. *farinacea* (Vain.) Vain. in Acta Soc. Fauna Fl. Fenn. 4: 339. 3-31 Dec 1887]. – Holotype: Chile, Magallanes, Puerto del Hambre ("Port Famine"), 1837-1840, *H. Jacquinot 16* (PC).

Cladonia fenestralis Nuno in J. Jap. Bot. 50: 291. Oct 1975. – Holotype: Malaysia, Sabah, Mt. Kinabalu, Paka Cave, 1975, *M. Togashi 66925* and in Kurokawa, Lich. Rar. Crit. Exsicc. No. 212 (TNS; isot.: B, H, UPS).

Cladonia fimbriata (L.) Fr., Lichenogr. Eur. Reform.: 222. Jun-Jul 1831. – Basionym: *Lichen fimbriatus* L., Sp. Pl.: 1152. 1 Mai 1753. – Lectotype (designated here): Dillenius, Hist. Musc.: t. 14, f. 8. 1742. Typotype: Herb. Dillenius (OXF).

Cladonia firma (Nyl.) Nyl. in Bot. Zeitung (Berlin) 19: 352. 22 Nov 1861. – Basionym: *Cladonia alcicornis* var. *firma* Nyl., Syn. Meth. Lich. 1: 191. Apr 1860. – Neotype (designated here): Portugal, Algarve, Marim, 1951, *C. N. Tavares* in Tavares, Lich. Lusit. Sel. Exsicc. No. 39 (H; isoneot.: LISU, UPS).

Cladonia floerkeana (Fr.) Flörke, De Cladon.: 99. Pasq. 1828. – Basionym: *Cenomyce floerkeana* Fr., Sched. Crit. Lich. Suec. 1-4: 18. 7 Mai 1825. – Lectotype (designated here): Sweden, *E. M. Fries* in Fries, Lich. Exsicc. Suec. No. 82 (UPS; isolectot.: B, FH-Tuck).

Cladonia floridana Vain. in Sandstede, Cladon. Exsicc. No. 1196. 22 Mai 1924. – Lectotype (designated here): U.S.A., Florida, Seminole Co.: Sanford, 1924, *S. Rapp* in Sandstede, Cladon. Exsicc. No. 1196 (TUR-V No. 15236; isolectot.: BM, H, FH, TUR-V No. 15240, UPS).

Cladonia foliacea (Huds.) Willd., Fl. Berol. Prodr.: 363. 1787. – Basionym: *Lichen foliaceus* Huds., Fl. Angl.: 457. Jan-Jun 1762. – Neotype (designated here): England, Salop (Shropshire), Haughmond Hill, *W. A. Leighton* in Leighton, Lich. Brit. Exsicc. No. 15 (BM; isoneot.: H, UPS).

Cladonia foliacea subsp. *convoluta* (Lam.) Badea & Cretz. in Acta Fauna Fl. Universali, Ser. 2, Bot. 1(15): 4. 1934. – Basionym and type: see under *Cladonia convoluta*.

Cladonia foliacea var. *convoluta* (Lam.) Vain. in Acta Soc. Fauna Fl. Fenn. 10: 394. Dec. 1894. – Basionym and type: see under *Cladonia convoluta*.

Cladonia fragilissima Østh. & P. James in Norweg. J. Bot. 24: 123. Jun 1977. – Holotype: Scotland, Main Argyll, Head of Loch Craigneish, on Allt Ath Mhic Mhàirtein, 1976, *P. W. James s.n.* (BM; isot.: E, H, O).

Cladonia friabilis Ahti in Lichenologist 22: 261. Jul 1990. – Holotype: Brazil, Bahia, Chapada Diamantina, Serra do Tombador, between Mundo Novo and Morro do Chapeú, 1980, *K. Kalb s.n.* (Herb. Kalb; isot.: H).

Cladonia fruticulosa Kremp. in Verh. K.K. Zool.-Bot. Ges. Wien 30, Abh.: 331. 1881. – Lectotype (Archer, 1986a): Australia, Queensland, Rockingham Bay, *J. Dallachy s.n.* ex Herb. F. von Mueller (M; isolectot.: MEL, NSW, REN-Abbayes).

Cladonia furcata (Huds.) Schrad., Spic. Fl. Germ.: 107. 16 Mai - 5 Jun 1794. – Basionym: *Lichen furcatus* Huds., Fl. Angl.: 458. Jan-Jun 1762. – Neotype (designated here): England, South Hunts, Winchfield, *W. A. Leighton* in Leighton, Lich. Brit. Exsicc. No. 401 (BM; isoneot.: H, UPS).

Cladonia furcata subsp. *subrangiformis* (Sandst.) Abbayes in Bull. Soc. Sci. Bretagne 14: 160, 162. 30 Mai 1938 ["1937"]. – Basionym and type: see under *Cladonia subrangiformis*.

Cladonia furfuracea Vain. in Acta Soc. Fauna Fl. Fenn. 10: 375. Dec 1894. – Lectotype (designated here): Brazil, Minas Gerais, Caraça, 1885, *E. A. Vainio* in Sandstede, Cladon. Exsicc. No. 1195 (TUR-V No. 19992; isolectot.: FH, G, H, TUR-V, UPS, US).

Cladonia fuscocinerea Ahti in Lichenologist 9: 7. Apr 1977. – Holotype: Uganda, Toro District, Ruwenzori Mts., Lower Bigo bog, Bigo Hut, 1966, *R. Cain & al. s.n.* (TRTC No. 66.3238; isot.: H, WIS).

Cladonia galindezii Øvstedal in Cryptog. Bryol. Lichénol. 9: 137. 1988. – Holotype: Antarctic, Graham Land, Argentine Islands, Galindez Island, 1935, British Graham Land (Penola) Expedition 1934-1937 No. 1108 (BM).

Cladonia gigantea (Bory) H. Olivier, Herb. Lich. Orne Calvados No. 401 [label]. 1884. – Basionym: *Lichen giganteus* Bory, Voy. Iles Afrique 3: 83. Sep 1804. – Lectotype (Ahti & Aptroot, 1992): Réunion ("Ile de Bourbon"), entrance of Coteau Maigre, 1801-1802, *J. Bory s.n.* (PC-Thur; isolectot.: PC, ?BM, ?G, ?H-ACH No. 1602, ?MB, ?PC).

Cladonia glauca Flörke, De Cladon.: 140. Pasq. 1828. – Lectotype (designated here): Germany, Brandenburg, Waren ("Wahren"), 1826, *H. G. Flörke 47* (H).

Cladonia glaucopallida Vain. in Nouv. Arch. Mus. Hist. Nat., ser. 3, 10: 267. 1898. – Holotype: Réunion, *Trapier 336* (TUR-V No. 15155; isot.: PC, PC-Hue).

Cladonia gracilenta Tuck. in Proc. Amer. Acad. Arts 5: 395. Mai 1862. – Lectotype (designated here): Cuba, Sagra de Savana, 1861, *C. Wright* in Wright, Lich. Ins. Cubae No. 43 (FH-Tuck; isolectot.: FH, G, H-NYL No. 37827, M, PC-Hue, TUR-V No. 14323, UPS).

Cladonia graciliformis Zahlbr. in Ann. Mycol. 14: 55. 1916. – Holotype: Japan, Mt. Sukawadake, *Aiba 2* (W).

Cladonia gracilis (L.) Willd., Fl. Berol. Prodr.: 363. 1787. – Basionym: *Lichen gracilis* L., Sp. Pl.: 1152. 1 Mai 1753. – Lectotype (designated here): Dillenius, Hist. Musc.: t. 14, f. 3. 1742. Typotype: Herb. Dillenius (OXF).

Cladonia gracilis subsp. ***elongata*** (Wulfen) Vain. in Acta Soc. Fauna Fl. Fenn. 53(1): 92. 4 Nov - 2 Dec 1922. – Basionym: *Lichen elongatus* Wulfen in Jacquin, Misc. Austriac. 2: 368. 1781. – Lectotype (Ahti, 1980): Chile, Magallanes, Baie de Bougainville, monts de Commerson, 1767, *P. Commerson s.n.* (PC; isolectot.: ?C, ?PC).

Cladonia gracilis subsp. ***tenerrima*** Ahti in Ann. Bot. Fenn. 17: 209. 4 Sep 1980. – Holotype: Australia, Victoria, Cathedral Range, North Jawbone, 5 km NW of Buxton, 1979, *R. Filson 16627* (MEL No. 1023710; isot.: BM, H).

Cladonia gracilis subsp. ***turbinata*** (Ach.) Ahti in Ann. Bot. Fenn. 17: 212. 4 Sep 1980. – Basionym: *Lichen turbinatus* Ach., Lichenogr. Suec. Prodr.: 192. 1799 ["1798"]. – Lectotype (Ahti, 1980): [Sweden?] (BM-Ach No. 702h; isolectot.: ?BM).

Cladonia gracilis subsp. ***valdiviensis*** Ahti in Ann. Bot. Fenn. 27: 322. 8 Jan 1991 ["1990"]. – Holotype: Chile, Malleco, Parque Nacional de Nahuelbuta, Pino Hueco, 1971, *M. Mahú 2433* (H; isot.: SGO).

Cladonia gracilis subsp. ***vulnerata*** Ahti in Ann. Bot. Fenn. 17: 207. 4 Sep 1980. – Holotype: U.S.A., Alaska, Chugach Mts., Blueberry Lake Campground (Richardson Highway Mile 23), 1967, *T. Ahti & J. W. Thomson 23642* (H; isot.: CANL, US, WIS).

Cladonia gracilis var. ***elongata*** (Wulfen) Fr., Lichenogr. Eur. Reform.: 219. Jun-Jul 1831. – Basionym and type: see under *Cladonia gracilis* subsp. *elongata*.

Cladonia gracilis var. ***turbinata*** (Ach.) Schaer., Enum. Crit. Lich. Eur.: 196. Aug-Sep 1850. – Basionym and type: see under *Cladonia gracilis* subsp. *turbinata*.

Cladonia granulans Vain. in Bot. Mag. (Tokyo) 35: 65. Mar-Apr 1921. – Holotype: Japan, Prov. Rikuzen, Mt. Katta, 1917, *A. Yasuda 124* (TUR-V No. 14174).

Cladonia granulosa (Vain.) Ahti in Ann. Bot. Fenn. 23: 205. 26 Nov 1986. – Basionym: *Cladonia subsquamosa* var. *granulosa* Vain. in Acta Soc. Fauna Fl. Fenn. 4: 448. 3-31 Dec 1887. – Lectotype (Ahti, 1986): Colombia, Antioquia, Sonsón, *G. Wallis s.n.* (TUR-V No. 15142; isolectot.: G).

Cladonia grayi G. Merr. ex Sandst., Cladon. Exsicc. No. 1847. 20 Mar 1929. – Lectotype (designated here): U.S.A., North Carolina ("N. Virg."), Mecklenburg Co., Charlotte, Long Creek, 1928, *F. W. Gray* in Sandstede, Cladon. Exsicc. No. 1847 (FH; isolectot.: B, FH, FH-Dodge, LE, S, TUR-V, UPS).

Cladonia groenlandica (Å. E. Dahl) Trass in Folia Cryptog. Estonica 1: 2. 15 Apr 1972. – Basionym and type: see under *Cladonia cornuta* var. *groenlandica*.

Cladonia guianensis Stenroos in Ann. Bot. Fenn. 26: 255. 31 Aug 1989. – Holotype: Venezuela, Bolívar, Distrito Piar, Macizo del Chimantá, between Torono-tepui and Chimantá-tepui, 1985, *T. Ahti & al. 45362* (VEN; isot.: H, NY, US).

Cladonia gymnopoda Vain. in Acta Soc. Fauna Fl. Fenn. 10: 172. Dec 1894. – Lectotype (designated here): Indonesia, Java, Mt. Magamendang, *G. Karsten 5* (ZT; isolectot.: TUR-V No. 18372).

Cladonia hedbergii Ahti in Lichenologist 9: 4. Apr 1977. – Holotype: Kenya, Central Province, Nanyuki District, Mt. Kenya, National Park Road (Naro Moru Track), 1970, *R. Santesson 22040* and in Moberg, Lich. Sel. Exsicc. Upsal. No. 56 [as *"32495"*] (UPS; isot.: CANL, H, O).

Cladonia hidakana Kurok. in Mem. Natl. Sci. Mus. 4: 61. Nov 1971. – Holotype: Japan, Hokkaido, Prov. Hidaka, Mt. Petegari, 1970, *S. Kurokawa 70310-b* (TNS).

Cladonia hokkaidensis Asahina in J. Jap. Bot. 44: 353. Dec 1969. – Holotype: Japan, Hokkaido, Prov. Ishikari, Mt. Daisetsu, Tomuraushi, 1969, *M. Togashi 69720* (TNS).

Cladonia homosekikaica Nuno in J. Jap. Bot. 50: 294. Oct 1975. – Holotype: Japan, Hokkaido, Prov. Nemuro (?), Shiretoko Peninsula, Mt. Rausu-dake, *M. Togashi 7075* (TNS).

Cladonia hondoënsis Asahina in J. Jap. Bot. 18: 551. Oct 1942. – Lectotype (designated here): Japan, Honshu, Prov. Shinano, Tsugaike, Mt. Shirouma, 1936, *Y. Asahina* in Asahina, Lich. Jap. Exsicc. No. 224 (TNS; isolectot.: H, UPS).

Cladonia humilis (With.) J. R. Laundon in Lichenologist 16: 220. Oct 1984. – Basionym: *Lichen humilis* With., Bot. Arr. Veg. Gr. Brit.: 721. 1776. – Holotype: Dillenius, Hist. Musc.: t. 14, f. 11. 1742. Typotype: England, London, Greenwich, Charlton and Woolwich, *J. J. Dillenius* (OXF).

Cladonia humilis var. *bourgeanica* A. W. Archer in Muelleria 7: 3. 12 Apr 1989. – Holotype: Australia, New South Wales, Six Foot Track, Binomea Ridge, 2 km N of Jenolan Caves, 1987, *A. W. Archer 2086* (MEL No. 1050873; isot.: NSW).

Cladonia hypoxantha Tuck. in Proc. Amer. Acad. Arts 5: 393. Mai 1862. – Lectotype (designated here): Cuba, Monte Verde, *C. Wright* in Wright, Lich. Ins. Cubae No. 41 (FH-Tuck; isolectot.: BM, M, PC, UPS).

Cladonia hypoxanthoides Vain. in Acta Soc. Fauna Fl. Fenn. 4: 135. 3-31 Dec 1887. – Lectotype (Stenroos, 1989b): Brazil, Minas

Gerais, Caraça, 1885, *E. A. Vainio s.n.* (TUR-V No. 14162).

Cladonia imbricarica Kristinsson in Lichenologist 6: 143. Oct 1974. – Holotype: Iceland, Central Highlands, Thjórsárver, Oddkelsalda, 1971, *H. Kristinsson 24582* (AMNH).

Cladonia imshaugii Ahti in Ann. Bot. Soc. Zool.-Bot. Fenn. "Vanamo" 32(1): 113. 25 Sep 1961. – Holotype: Dominian Republic, La Vega, Cordillera Central, trail to Alto de la Bandera, ravine with stream entering Valle de Los Robles, 1958, *H. A. Imshaug 23493* (H; isot.: MSC).

Cladonia incrassata Flörke, De Cladon.: 21. Pasq. 1828. – Lectotype (designated here): Germany, *H. G. Flörke* in Flörke, Cladon. Exsicc. No. 5 (G).

Cladonia inobeana Asahina in J. Jap. Bot. 38: 1. Jan 1963. – Holotype: Japan, Shikoku, Prov. Awa, Kaibu-gun, Kainan-machi, 1962, *T. Inobe 65* (TNS).

Cladonia insignis Nyl. in Ann. Sci. Nat., Bot., ser. 4, 11: 250. Apr 1859. – Lectotype (Stenroos, 1991): Réunion, Salazié, 1840, *P. Lepervanche-Mézières 112* (PC-Thur; isolectot.: G, H-NYL No. 37844).

Cladonia insolita Ahti & Krog in Ann. Bot. Fenn. 24: 85. 30 Jun 1987. – Holotype: Uganda, Ruwenzori, Mt. Stanley, 1948, *O. Hedberg 724i* (UPS).

Cladonia intermediella Vain. in Acta Soc. Fauna Fl. Fenn. 10: 12.

Dec 1894. – Holotype: Mauritius, Mt. Pouce, *P. B. Ayres s.n.* (BM).

Cladonia isabellina Vain. in Acta Soc. Fauna Fl. Fenn. 10: 174. Dec 1894. – Holotype: Colombia ("Nova Granata"), San Isabel, *G. Wallis s.n.* (G; isot.: G, S, TUR-V No. 18382).

Cladonia japonica Vain. in Nouv. Arch. Mus. Hist. Nat., ser. 3, 10: 265. 1898. – Lectotype (designated here): Japan, Prov. Oshima, Hakodate ("Hakkoda"!), 1897, *R. P. Faurie 369 p.p.* (TUR-V No. 15160; isolectot.: PC).

Cladonia kanewskii Oxner in Ukrajins'k. Bot. Žurn. 3: 9. 1926. – Holotype: Russia, Buryatia, Barguzin District, Svyatoy Nos by Lake Baikal, Barmashev Lake, 1916, *G. Kanevskiy s.n.* (KW; isot.: LE).

Cladonia kauaiensis H. Magn. & Zahlbr. in Ark. Bot. 31A(6): 28. 21 Apr 1944. – Lectotype (Stenroos, 1986): U.S.A., Hawaii, Kauai, Kaholuamanoa, above Waimea, 1895, *A. A. Heller 2797* and in Merrill, Lich. Exsicc. No. 89 (UPS; isolectot.: B, CANL, FH, G, H, TUR-V No. 14172).

Cladonia koyaënsis Asahina in J. Jap. Bot. 28: 162. Jun 1953. – Holotype: Japan, Honshu, Prov. Kii, Mt. Koya, 1952, *Y. Asahina 2774* (TNS; isot.: US).

Cladonia krempelhuberi (Vain.) Zahlbr., Cat. Lich. Univ. 4: 552. Feb 1927. – Basionym: *Cladonia verticillata* var. *krempelhuberi* Vain. in Acta Soc. Fauna Fl. Fenn.

10: 187. Dec 1894. – Lectotype (designated here): New Zealand, *C. Knight s.n.* (M; isolectot.: TUR-V No. 18318).

Cladonia kuringaiensis A. W. Archer in Muelleria 4: 273. 26 Mai 1980. – Holotype: Australia, New South Wales, Ku-ring-gai Chase National Park, Spring Gully Creek, 4 km SSW of Bobbin Head, 1979, *A. W. Archer 620* (MEL No. 1023704; isot.: NSW).

Cladonia labradorica Ahti & Brodo in Bryologist 84: 238. 31 Aug 1981. – Holotype: Canada, Québec, Nouveau-Québec, Poste-de-la-Baleine, 1974, *A. Vachon s.n.* (CANL; isot.: H).

Cladonia laevigata (Vain.) Gyeln., Lichenotheca No. 148 [label]. 1937. – Basionym: *Cladonia sylvatica* var. *laevigata* Vain. in J. Bot. (Morot) 1: 284. 1 Nov 1887 [and in Acta Soc. Fauna Fl. Fenn. 4: 33. 3-31 Dec 1887]. – Holotype: Chile, Antártica Chilena, Isla Cabo de Hornos ("Cap Horn"), Bahía Orange, 1883, *P. Hariot s.n.* (TUR-V No. 13335; isot.: G, PC, UPS).

Cladonia laii Stenroos in Acta Bryolichenol. Asiat. 1: 53. 17 Apr 1990 ["1989"]. – Holotype: China, Taiwan, Chiayi-Nantou, Yu-shan National Park, between Pai-yun and Mt. Morrison, 1987, *S. Stenroos 3225* (H; isot.: Herb. Lai).

Cladonia leiodea H. Magn. in Ark. Bot. 30B(3): 1. 25 Feb 1942. – Holotype: U.S.A., Hawaii, Kauai, Alakai, Kilohana, 1938, *O. Selling 5674* (S).

Cladonia lepidophora Ahti & Kashiw. in Inoue, Stud. Cryptog. S. Chile: 140. 30 Aug 1984. – Holotype: Chile, Llanquihue, Volcán Osorno, 1981, *H. Kashiwadani 17106* (TNS; isot.: H, SGO).

Cladonia leporina Fr., Lichenogr. Eur. Reform.: 243. Jun-Jul 1831. – Lectotype (designated here): U.S.A., North Carolina, Forsyth Co., Salem, *L. von Schweinitz s.n.* (UPS).

Cladonia leprocephala Ahti & Stenroos in Ann. Bot. Fenn. 23: 236. 26 Nov 1986. – Holotype: Venezuela, Mérida, Páramo La Negra, between Delgadito and Portachuelo, 1974, *M. E. Hale 42481* (US; isot.: H, MERF).

Cladonia leucophylla Ahti & Krog in Ann. Bot. Fenn. 24: 86. 30 Jun 1987. – Holotype: Kenya, Central Province, Kirinyaga District, Mt. Kenya, Castle Forest Station, 1977, *T. D. V. Swinscow 5k 7/16* (BM; isot.: O).

Cladonia libifera Savicz in Novosti Sist. Nizš. Rast. 1965: 167. 30 Apr 1965. – Holotype: Russia, Yakutia, Aldan River, Chandyga, 1949, *L. N. Tjulina s.n.* and in Savicz, Lichenoth. Ross. No. 150 (LE; isot.: BM, CANL, FH, H, UPS).

Cladonia lopezii Stenroos in Ann. Bot. Fenn. 26: 250. 31 Aug 1989. – Holotype: Venezuela, Mérida, Distrito Libertador, Parque Nacional Sierra Nevada de Mérida, between La Aguada and La

Montaña, 1981, *M. Lindström 875* (GB; isot.: H, MERF).

Cladonia luteoalba Wheldon & A. Wilson, Fl. W. Lancashire: 450. Sep 1907 [and A. Wilson & Wheldon in Trans. Liverpool Bot. Soc. 1: 6. 1909]. – Lectotype (Ahti, 1965): England, West Lancashire, Greygarth Fall, 1906, *A. Wheldon & A. Wilson 19* (BM; isolectot.: NMW).

Cladonia luzonensis Ahti in Ann. Bot. Soc. Zool.-Bot. Fenn. "Vanamo" 32(1): 130. 25 Sep 1961. – Holotype: The Philippines, Luzon, Prov. Benguet, Mt. Pulog, 1909, *H. M. Curran & al.* in Philippines Forestry Bureau No. 16374 (TUR-V No. 12910; isot.: US, W).

Cladonia macilenta Hoffm., Deutschl. Fl. 2: 126. 1796. – Neotype (designated here): Germany, Niedersachsen, Oldenburg, Litteler Fuhrenkamp, 1919, *H. Sandstede* in Sandstede, Cladon. Exsicc. No. 477 (UPS; isoneot.: FH, UPS).

Cladonia macilenta subsp. *bacillaris* (Genth) Boistel, Nouv. Fl. Lich. 2: 24. 1902. – Basionym and type: see under *Cladonia bacillaris.*

Cladonia macilenta var. *bacillaris* (Genth) Schaer., Enum. Crit. Lich. Eur.: 186. Aug-Sep 1850. – Basionym and type: see under *Cladonia bacillaris.*

Cladonia macroceras (Delise) Hav. in Bergens Mus. Årbok, Naturvidensk. Rekke, 1927(3): 12. 1927. – Basionym: *Cenomyce gracilis* var. *macroceras* Delise in Duby, Bot. Gall.: 624. Mai 1830. – Type not designated.

Cladonia macrophylla (Schaer.) Stenh., Lich. Suec. Exsicc., ed. 2, No. 186 [label]. 1865 [and in Öfvers. Förh. Kongl. Svenska Vetensk.-Akad. 22: 231. 1865]. – Basionym: *Cladonia ventricosa* var. *macrophylla* Schaer., Lich. Helv. Spic.: 316. 1833. – Lectotype (designated here): Switzerland, Canton Berne, Grimsel Pass, Handeck; and Susten Pass, Gadmen; *L. E. Schaerer* in Schaerer, Lich. Helv. Exsicc. No. 279 (G; isolectot.: BERN, G, H, UPS).

Cladonia macrophylliza Vain. in Acta Soc. Fauna Fl. Fenn. 10: 7. Dec 1894. – Lectotype (designated here): Cuba, Filantropia, 1857, *C. Wright* in Wright, Lich. Ins. Cubae No. 26 (TUR-V No. 17229; isolectot.: BM, FH, FH-Tuck, G, H-NYL, M, UPS, US).

Cladonia macrophyllodes Nyl. in Flora 58: 447. 1 Oct 1875. – Lectotype (designated here): Austria, Tyrol, Kühtai ("Kühthei"), 1874, *F. Arnold 23* (H-NYL No. 38748).

Cladonia macroptera Räsänen in J. Jap. Bot. 16: 149. Mar 1940. – Holotype: Japan, Prov. Shinano, Mt. Yatsugatake, 1918, *A. Yasuda 498* (H; isolectot.: H, TNS).

Cladonia magnussonii Ahti in Ann. Bot. Soc. Zool.-Bot. Fenn. "Vanamo" 32(1): 99. 25 Sep 1961. – Holotype: U.S.A., Hawaii, Maui, Puu Kukui, 1922, *C. Skottsberg 1359* (GB).

Cladonia magyarica Vain. in Fl. Hung. Exsicc. No. 715 [label]. 1927 [and in Sched. Fl. Hung. Exsicc. 8: 8. 1927]. – Lectotype (Verseghy, 1964): Hungary, Pest, forest Bugaci nagyerdö near Kecskemét, 1923, *G. Timkó* in Fl. Hung. Exsicc. No. 715 (BP No. 13365; isolectot.: B, BM, CANL, FH, H, S, TUR-V No. 12570, UPS, US).

Cladonia mascarena Nyl. in Ann. Sci. Nat., Bot., ser. 4, 11: 250. Apr 1859. – Lectotype (designated here): Réunion, 1840, *P. Leper-vanche-Mézières 116* (H-NYL No. 38692; isolectot.: G, PC-Hue, PC-Thur).

Cladonia mateocyatha Robbins in Rhodora 27: 50. 28 Apr 1925. – Lectotype (designated here): U.S.A., Massachusetts, Plymouth Co., Wareham, *C. A. Robbins s.n.* (FH).

Cladonia mauiensis H. Magn. in Ark. Bot. 30B(3): 3. 25 Feb 1942. – Holotype: U.S.A., Maui, Halea-kala, Kula pipe line, 1929, *C. Skottsberg 1167* (S; isot.: UPS).

Cladonia mawsonii C. W. Dodge in Brit. Austral. New Zealand Ant-arct. Res. Exped. 1929-1931 Rep., Ser. B, 7: 128. 1948. – Type not designated.

Cladonia maxima (Asahina) Ahti, Ann. Bot. Fenn. 15: 12. 8 Mai 1978. – Basionym: *Cladonia gra-cilis* f. *maxima* Asahina, Atlas Jap. Cladon.: 19, f. 95. 1971. – Holo-type: Japan, Hokkaido, Abashiri District, Shiretoko Peninsula, Mt.

Rausu, 1970, *M. Togashi s.n.* (TNS).

Cladonia mediterranea P. A. Du-vign. & Abbayes in Rev. Bryol. Lichénol. 16: 95. 1947. – Holo-type: France, Gard, Bellegarde, mas de Broussan, 1946, *P. A. Du-vigneaud s.n.* (PC; isot.: BRLU, FH, UPS).

Cladonia medusina (Bory) Nyl. in Mém. Soc. Sci. Nat. Cherbourg 5: 95. 24-30 Mar 1858 ["1857"]. – Basionym: *Lichen medusinus* Bo-ry, Voy. Iles Afrique 3: 102. Sep 1804. – Lectotype (Huovinen & Ahti, 1986): Réunion, Les hauts de la Plaine des Cafres, en alant au Coteau Maigre, *J. Bory s.n.* (PC-Thur; isolectot.: PC).

Cladonia melaleuca Nuno in J. Jap. Bot. 47: 162. Jun 1972. – Holo-type: Malaysia, Pahang, Cameron Highlands, Gunon Bringchang, 1970, *M. Togashi 70305* (TNS).

Cladonia melanocaulis Stenroos in Ann. Bot. Fenn. 25: 138. 9 Aug 1988. – Holotype: Papua New Guinea, Central District, Mt. Al-bert-Edward, en route from the Woitape Airstrip to summit of Mt. Albert-Edward, 1974, *H. Kashi-wadani 12299* (TNS; isot.: H).

Cladonia meridensis Ahti & Sten-roos in Ann. Bot. Fenn. 23: 234. 26 Nov 1986. – Holotype: Vene-zuela, Táchira, Distrito Jáuregui, Mun. Vargas, Páramo de El Zum-bador, c. 5 km S of El Cobre, 1985, *T. Ahti & M. López-Fi-gueiras 43828* (MERF; isot.: H, US, VEN).

Cladonia meridionalis Vain. in Denkschr. Kaiserl. Akad. Wiss., Wien, Math.- Naturwiss. Kl., 83: 136. Nov-Dec 1909. – Lectotype (designated here): Brazil, São Paulo, Raiz da Serra, 1901, *V. Schiffner 9* (TUR-V No. 14175; isolectot.: US).

Cladonia merochlorophaea Asahina in J. Jap. Bot. 16: 713. 15 Dec 1940. – Lectotype (designated here): Germany, Niedersachsen, Oldenburg, Oldenburger Sand, 1918, *H. Sandstede* in Sandstede, Cladon. Exsicc. No. 389 (TNS; isolectot.: B, FH, H, UPS).

Cladonia merochlorophaea var. *novochlorophaea* Sipman in Acta Bot. Neerl. 22: 496. 12 Nov 1973. – Holotype: The Netherlands, Terschelling Island, Boschplaat, 1971, *H. J. M. Sipman 4895* (U).

Cladonia metacorallifera Asahina in J. Jap. Bot. 15: 612. Oct 1939. – Lectotype (Tønsberg, 1975): Japan, Honshu, Prov. Shinano, Mt. Norikura, Katanokoya Lodge, 1939, *Y. Asahina 3144* (TNS).

Cladonia metacorallifera var. *reagens* Asahina in J. Jap. Bot. 15: 615. Oct 1939. – Lectotype (Stenroos, 1989b): Japan, Honshu, Prov. Etchu, Mt. Yakushi, 1936, *Y. Asahina 36725* (TNS).

Cladonia mexicana Vain. in Acta Soc. Fauna Fl. Fenn. 4: 452. 3-31 Dec 1887. – Lectotype (designated here): Mexico, San Luis Potosí, 1851, *M. Virlet d'Aoust* in Herb. E. Fournier No. 97 (PC).

Cladonia microscypha Ahti & Stenroos in Ann. Bot. Fenn. 23: 232. 26 Nov 1986. – Holotype: Venezuela, Mérida, Sierra Nevada de Santo Domingo, Páramo de Mucuchies, séctor Mucubají, between Laguna Negra and Mucubají, 1975, *M. E. Hale & M. López-Figueiras 44447* (US; isot.: H, MERF).

Cladonia minarum Ahti in Ann. Bot. Fenn. 23: 212. 26 Nov 1986. – Holotype: Brazil, Minas Gerais, Caraça, 1885, *E. A. Vainio s.n.* (TUR-V No. 13617).

Cladonia miniata G. Mey., Nebenst. Beschaeft. Pflanzenk.: 149. Jul-Dec 1825. – Neotype (Stenroos, 1989a): Brazil, Rio de Janeiro, Mun. Resende, Itatiaya, 5 km ENE of Alto da Serra, 1987, *T. Ahti & P. G. Windisch 45975* (SP; isoneot.: H, NY).

Cladonia miniata var. *anaemica* (Nyl.) Zahlbr., Cat. Lich. Univ. 4: 561. Feb 1927. – Basionym: *Cladonia sanguinea* var. *anaemica* Nyl., Syn. Meth. Lich. 1: 219. Apr 1860. – Lectotype (Stenroos, 1989a): Brazil (H-NYL No. 37851).

Cladonia miniata var. *parvipes* (Vain.) Zahlbr., Cat. Lich. Univ. 4: 561. Feb 1927. – Basionym: *Cladonia miniata* f. *parvipes* Vain. in Acta Soc. Fauna Fl. Fenn. 4: 66. 3-31 Dec 1887. – Lectotype (Stenroos, 1989a): Brazil, Minas Gerais, Caraça, 1885, *E. A. Vainio s.n.* (TUR-V No. 14189).

Cladonia mitis Sandst., Cladon. Exsicc. No. 55. 20 Mar 1918. –

Lectotype (Ruoss, 1987): Germany, Niedersachsen, Oldenburg, Kronsberge near Bösel, 1916, *H. Sandstede* in Sandstede, Cladon. Exsicc. No. 55 (BRNU; isolectot.: B, BERN, BRNU, FH, H, NMLU, TUR-V No. 13084, UPS).

Cladonia modesta Ahti & Krog in Ann. Bot. Fenn. 24: 88. 30 Jun 1987. – Holotype: Kenya, Central Province, Kirinyaga District, Mt. Kenya, near Castle Forest Station, 1972, *H. Krog & T. D. V. Swinscow K 49/156* (O).

Cladonia mongolica Ahti in Nova Hedwigia 44: 196. Feb 1987. – Holotype: Mongolia, Chövsgöl Aimak, Ulan-Uul Somon, Belmes-Gol valley, S of Tomin brigade, 1983, *S. Huneck MVR-83-80* (H; isot.: Herb. Huneck).

Cladonia multiformis G. Merr. in Bryologist 12: 1. Jan 1909. – Type not designated.

Cladonia murrayi W. Martin in Trans. Roy. Soc. New Zealand, Bot. 2: 40. 30 Nov 1962. – Holotype: New Zealand, West Otago, Doubtful Sound, Secretary Island, 1959, *J. Murray s.n.* (CHR No. 257075; isot.: BM).

Cladonia mutabilis Vain. in Acta Soc. Fauna Fl. Fenn. 4: 297. 3-31 Dec 1887. – Lectotype (designated here): Brazil, Minas Gerais, Caraça, 1885, *E. A. Vainio s.n.* (TUR-V No. 15230; isolectot.: US).

Cladonia nana Vain. in Acta Soc. Fauna Fl. Fenn. 10: 23. Dec 1894. – Lectotype (designated here): Brazil, Minas Gerais, Antônio Carlos (Sitio), 1885, *E. A. Vainio s.n.* (TUR-V No. 17230).

Cladonia neozelandica Vain. in Acta Soc. Fauna Fl. Fenn. 10: 34. Dec 1894. – Holotype: New Zealand (probably Wellington), *C. Knight s.n.* (M; isot.: TUR-V No. 17286) [photograph in Arnold, Lich. Exsicc. No. 1643 (G, H)].

Cladonia nipponica Asahina, Lich. Japan 1: 123. 1950. – Lectotype (Ahti, 1973): Japan, Honshu, Prov. Kai, Sensui Pass, 1937, *Y. Asahina 37014* (TNS).

Cladonia nitida Ahti in Mycosystema 4: 61. 17 Jan 1992 ["1991"]. – Holotype: Vietnam ("Tonkin"), Vodivato, 1889, *R.-P. Bon 1703* (PC-Hue).

Cladonia norikurensis Asahina in J. Jap. Bot. 28: 121. Apr 1953. – Holotype: Japan, Honshu, Prov. Shinano, Mt. Norikura, Suzuran, *Y. Asahina s.n.* (TNS).

Cladonia norvegica Tønsberg & Holien in Nordic J. Bot. 4: 79. 1984. – Holotype: Norway, Sør-Trøndelag, Melhus, 1982, *T. Tønsberg 6870* and in Vězda, Lich. Sel. Exsicc. No. 1978 (TRH; isot.: B, BM, CANL, H, S, UPS, US).

Cladonia oceanica (Vain.) Zahlbr., Lich. Rar. Exsicc. No. 158. Mai 1911. – Basionym: *Cladonia didyma* subsp. *oceanica* Vain. in Acta Soc. Fauna Fl. Fenn. 4: 147. 3-31 Dec 1887. – Lectotype (Stenroos, 1986): U.S.A., Hawaii, Kauai, 1851-1855, *J. Rémy* (PC).

Cladonia ochracea L. Scriba in Sandstede, Cladon. Exsicc. No. 1006. 1923. – Lectotype (designated here): Brazil, Rio Grande do Sul, Porto Alegre, 1907, *A. Stier* in Sandstede, Cladon. Exsicc. No. 1006 (H; isolectot.: B, BM, FH, G, M, UPS, US).

Cladonia ochrochlora Flörke, De Cladon.: 75. Pasq. 1828. – Neotype (designated here): Germany, Niedersachsen, Oldenburg, Oldenburger Sand, 1918, *H. Sandstede* in Sandstede, Cladon. Exsicc. No. 241 (UPS; isoneot.: FH, H).

Cladonia pachycladodes Vain. in Sandstede, Cladon. Exsicc. No. 1141. 22 Mai 1924. – Lectotype (Ahti, 1973): U.S.A., Florida, Seminole Co., Sanford, 1923, *S. Rapp* in Sandstede, Cladon. Exsicc. No. 1141 (TUR-V No. 13556; isolectot.: FH, H, TUR-V, UPS).

Cladonia pachyclados (Vain.) Ahti in Ann. Bot. Fenn. 23: 213. 26 Nov 1986. – Basionym: *Cladonia pycnoclada* f. *pachyclados* Vain. in Nouv. Arch. Mus. Hist. Nat., ser. 3, 10: 258. 1898. – Type not designated.

Cladonia paeminosa A. W. Archer in Muelleria 7: 1. 12 Apr 1989. – Holotype: Australia, Victoria, near Mirimbah, Mt. Stirling Road, c. 30 km E of Mansfield, 1986, *A. W. Archer 2027* (MEL No. 1050876; isot.: ANUC, CBG, H, NSW).

Cladonia pallens Ahti & Krog in Ann. Bot. Fenn. 24: 89. 30 Jun 1987. – Holotype: Tanzania, Northern Province, Arusha District, Mt. Meru Crater, 1974, *H. Krog & T. D. V. Swinscow T 5/198* (O; isot.: BM, H).

Cladonia papuana Stenroos in Ann. Bot. Fenn. 23: 161. 26 Jun 1986. – Holotype: Papua New Guinea, Morobe Province, Mt. Sarawaket Southern Range, 2.5 km S of Lake Gwam, Lake Kenzoroh, 1981, *T. Koponen 32604* (H; isot.: LAE).

Cladonia parasitica (Hoffm.) Hoffm., Deutschl. Fl. 2: 127. 1796. – Basionym: *Lichen parasiticus* Hoffm., Enum. Lich.: 39. 1784. – Neotype (designated here): Germany, Bavaria, Munich, W of Deisenhofen, Grünwald, 1892, *F. Arnold* in Rehm, Cladon. Exsicc. No. 410 (M; isoneot.: H, UPS).

Cladonia parva Ahti & Krog in Ann. Bot. Fenn. 24: 89. 30 Jun 1987. – Holotype: Tanzania, Eastern Province, Morogoro District, Mindu Forest Reserve, 1978, *Å. E. Dahl s.n.* (O).

Cladonia patagonica A. Evans in Rev. Bryol. Lichénol. 24: 135. Jan-Aug 1955. – Holotype: Argentina, Chubut, Los Alerces National Park, Lago Menéndez, 1950, *I. M. Lamb 5886* (US; isot.: BM, CANL, SI, UPS).

Cladonia peltasta Spreng., Syst. Veg. 4(1): 271. 1-7 Jun 1827. – Lectotype (Ahti & Aptroot, 1992): Réunion, Plaine des Chicots "et les hauts du Bras de la Plaine", *J. Bory*, "à Willdenow 50 et à Flörke 15" (PC-Thur; isolectot.: BM).

Cladonia peltastica (Nyl.) Müller Arg. in Flora 63: 260. 11 Jun 1880. – Basionym: *Cladina peltastica* Nyl., Flora 57: 70. 11 Feb 1874. – Lectotype (Huovinen & Ahti, 1986): Brazil, Amazonas, Umirisál opposite to Manaus at mouth of Rio Negro, 1849-1855, *R. Spruce* in Spruce, Lich. Amaz. And. No. 22 (H-NYL No. 37612; isolectot.: BM, G, NY).

Cladonia perfilata Hook. in Icon. Pl. 2: t. 192. 7 Oct - 8 Nov 1837. – Holotype: Brazil, Rio de Janeiro, Serra dos Orgãos, Tejuca, Pedra Bonita, 1836, *G. Gardner 3155* (BM; isot.: BM, M).

Cladonia perforata A. Evans in Trans. Connecticut Acad. Arts 38: 326. Sep 1952. – Holotype: U.S.A., Florida, Escambia Co., Santa Rosa Island, 1945, *G. A. Llano s.n.* (US; isot.: FH).

Cladonia perfossa Nuno in J. Jap. Bot. 47: 164. Jun 1972. – Holotype: China, Taiwan, Prov. Chiayi, Mt. Morrison (Shi-kaosan), Mt. Shi-san, 1964, *S. Kurokawa 440* (TNS).

Cladonia perlomera Kristinsson in Bryologist 72: 432. 4 Feb 1970. – Holotype: U.S.A., North Carolina, Pender Co., 4 mi. W of Hampstead, 1969, *H. Kristinsson 30150* (DUKE; isot.: US).

Cladonia perplexa Abbayes in Rev. Bryol. Lichénol. 16: 88. 1947. – Lectotype (Stenroos, 1991): Madagascar, Toliara, Taolanaro

(Fort-Dauphin), 1932, *R. Decary s.n.* (PC).

Cladonia perrieri Abbayes in Rev. Bryol. Lichénol. 16: 80. 1947. – Lectotype (designated here): Madagascar, Massif d'Andringitra, 1911, *J. H. Perrier de la Bathie s.n.* (PC; isolectot.: REN-Abbayes).

Cladonia pertricosa Kremp. in Verh. K.K. Zool.-Bot. Ges. Wien 30, Abh.: 331. 1881. – Holotype: Australia (M; isot.: MEL No. 6619).

Cladonia petrophila R. C. Harris in Brittonia 44: 326. 28 Sep 1992. – Holotype: U.S.A., Tennessee, Carter Co., Cherokee National Forest, Dennis Cove, 1976, *R. C. Harris 10730* (NY).

Cladonia peziziformis (With.) J. R. Laundon in Lichenologist 16: 223. Oct 1984. – Basionym: *Lichen peziziformis* With., Bot. Arr. Veg. Gr. Brit.: 720. 1776. – Holotype: Dillenius, Hist. Musc.: t. 14, f. 2. 1742. Typotype: Herb. Sherard No. 1774 (OXF).

Cladonia phyllophora Hoffm., Deutschl. Fl. 2: 123. 1796. – Lectotype (designated here): Germany, Niedersachsen, Hannover, Herrenhausen, *F. Ehrhart* in Ehrhart, Pl. Crypt. No. 287 (GOET; isolectot.: LINN-Sm No. 1709.1).

Cladonia phyllopoda (Vain.) Stenroos in Ann. Bot. Fenn. 25: 132. 9 Aug 1988. – Basionym: *Cladonia pityrea* var. *phyllopoda* Vain. in Ann. Acad. Sci. Fenn., Ser. A, 15(6): 55. 1921. – Holotype: The

Philippines, Luzon, Ifugao, Mt. Polis, 1913, *R. C. McGregor* in Bur. Sci. No. 20354 p.p. (TUR-V No. 19743; isot.: BM).

Cladonia physodalica Elix, Lich. Australas. Exsicc. 9: [2]. Dec 1990. – Holotype: Australia, Queensland, Main Coast Range, Mt. Lewis Track, Mary Creek, 15 km NNW of Mt. Mollow, 1984, *J. A. Elix & H. Streimann 16947* in Elix, Lich. Australas. Exsicc. No. 204 (CBG; isot.: FH, H, UPS).

Cladonia piedmontensis G. Merr. in Bryologist 27: 22. 6 Mai 1924. – Lectotype (designated here): U.S.A., North Carolina, Mecklenburg Co., Charlotte, 1924, *F. W. Gray 288* (FH; isolectot.: FH).

Cladonia pityrophylla Nyl. in Flora 57: 70. 11 Feb 1874. – Holotype: Brazil, Pará, Santarém, mouth of Rio Tapajos, 1849-1855, *R. Spruce* in Spruce, Lich. Amaz. And. No. 26 (H-NYL No. 38843; isot.: BM, G, PC, TUR-V No. 19989).

Cladonia pleurophylla Vain. in Acta Soc. Fauna Fl. Fenn. 4: 508. 3-31 Dec 1887. – Lectotype (designated here): Brazil, Minas Gerais, Caraça, 1885, *E. A. Vainio s.n.* (TUR-V No. 19991; isolectot.: TUR-V).

Cladonia pleurota (Flörke) Schaer., Enum. Crit. Lich. Eur.: 186. Aug-Sep 1850. – Basionym: *Capitularia pleurota* Flörke in Ges. Naturf. Freunde Berlin Mag. Neuesten Entdeck. Gesammten Naturk. 2: 217. 1808. – Neotype (designated here): Germany or Austria, "data

a Flörke Berolini 1811" (UPS; isoneot.: ?BM, ?H).

Cladonia pocillum (Ach.) Grognot, Pl. Crypt. Saône-et-Loire: 82. 1863. – Basionym: *Baeomyces pocillum* Ach., Methodus: 336. Jan-Apr 1803. – Lectotype (Ahti, 1966): Sweden (H-ACH No. 1656A; isolectot.: BM).

Cladonia poeciloclada Abbayes in Rev. Bryol. Lichénol. 33: 236. 1964. – Holotype: South Africa, Cape Province, District Knysna, Tsitzikama Mts., Platt Forest near Stormsriver, 1949, *R. A. Maas Geesteranus 6647* (L; isot.: H, REN-Abbayes).

Cladonia polycarpia G. Merr. in Bryologist 12: 46. Mai 1909. – Lectotype (Evans, 1944): U.S.A., South Carolina, Beaufort Co.: Beaufort, 1868, *J. H. Mellichamp s.n.* (FH).

Cladonia polycarpoides Nyl. in Zwackh, Lich. Exsicc. No. 626, 626bis [correction labels]. 1892. – Lectotype (Suominen & Ahti, 1966): Switzerland, Zürich, Ober-Riffersweil, 1884, *C. Hegetschweiler* in Zwackh, Lich. Exsicc. No. 626bis; and in Rehm, Cladon. Exsicc. No. 315; and in Lojka, Lichenoth. Univ. No. 3 (H-NYL No. p.m. 963; isolectot.: BM, H, TUR-V No. 17421, UPS).

Cladonia polydactyla (Flörke) Spreng., Syst. Veg. 4(1): 274. 1-7 Jun 1827. – Basionym: *Cenomyce polydactyla* Flörke, Deutsche Lich. 10: 13. 1821. – Lectotype (designated here): Germany,

Rostock, *H. G. Flörke* in Flörke, Deutsche Lich. No. 195A (UPS).

Cladonia polytypa Vain. in Acta Soc. Fauna Fl. Fenn. 4: 301. 3-31 Dec 1887. – Lectotype (designated here): Brazil, Minas Gerais, Caraça, 1885, *E. A. Vainio s.n.* (TUR-V No. 15243; isolectot.: TUR-V).

Cladonia poroscypha S. Hammer in Bryologist 96: 83. 22 Mar 1993. – Holotype: U.S.A., California, San Mateo Co., Pilarcitos Creek Canyon, 1943, *Herre 3* (US).

Cladonia portentosa (Dufour) Coem. in Bull. Acad. Roy. Sci. Belgique, ser. 2, 19: 43, 49. 1865. – Basionym: *Cenomyce portentosa* Dufour, Ann. Gén. Sci. Phys. 8: 69. Mai 1821. – Lectotype (Ahti, 1978): France, Landes, St-Sever, *J. Dufour s.n.* (PC-Desm; isolectot.: H-ACH No. 1952, TUR-V No. 13303).

Cladonia portentosa subsp. ***pacifica*** (Ahti) Ahti in Ann. Bot. Fenn. 15: 9. 8 Mai 1978. – Basionym: *Cladonia pacifica* Ahti in Ann. Bot. Soc. Zool.-Bot. Fenn. "Vanamo" 32(1): 79. 25 Sep 1961. – Holotype: Canada, British Columbia, West Vancouver, Horseshoe Bay, Whytecliffe, 1958, *V. J. Krajina s.n.* (H; isot.: UBC).

Cladonia praetermissa A. W. Archer in Muelleria 5: 273. 5 Apr 1984. – Holotype: Australia, New South Wales, Epping, Devlin's Creek, 1982, *A. W. Archer 1376* (MEL No. 1036220; isot.: H, NSW).

Cladonia praetermissa var. ***modesta*** (Ahti & Krog) Kantvilas & A. W. Archer in Nordic J. Bot. 11: 369. Aug 1991. – Basionym and type: see under *Cladonia modesta.*

Cladonia prolifica Ahti & S. Hammer in Mycotaxon 37: 342. 23 Apr 1990. – Holotype: U.S.A., California, Amador Co., Ione, Old Lambert Road, 1988, *S. Hammer 1265* (FH; isot.: H, NY).

Cladonia prostrata A. Evans in Trans. Connecticut Acad. Arts 38: 281. Sep 1952. – Holotype: U.S.A., Florida, Lake Co., Eustis, 1947, *A. Evans 170* (US).

Cladonia pseudalcicornis Asahina in J. Jap. Bot. 19: 192. Jul 1943. – Type not designated.

Cladonia pseudodidyma Asahina in J. Jap. Bot. 15: 667. Nov 1939. – Lectotype (designated here): Japan, Honshu, Prov. Musashi, Mt. Bukosan, 1922, *Y. Asahina 270* (TNS).

Cladonia pseudoëvansii Asahina in J. Jap. Bot. 16: 187. 15 Apr 1940. – Lectotype (Ahti, 1961): Japan, 1923, *M. Hiroe s.n.* (TNS; isolectot.: H).

Cladonia pseudogymnopoda Asahina in J. Jap. Bot. 45: 260. Sep 1970. – Holotype: Japan, Shikoku, Prov. Awa, Kaibu-gun, Kainan-machi, Karei-dani (Todoroki Shrine), 1955, *M. Togashi 55725* (TNS).

Cladonia pseudohondoënsis Asahina in J. Jap. Bot. 34: 348. Nov 1959. – Holotype: Japan, Prov.

Ettyu, Mt. Taro, 1936, *Y. Asahina s.n.* (TNS).

Cladonia pseudomacilenta Asahina in J. Jap. Bot. 19: 190. Jul 1943. – Holotype: Japan, Honshu, Prov. Shimotsuke, Nikko, Chuzenji Lake, Ase-wan, 1942, *S. Kato* ex Herb. Asahina No. 13198 (TNS; isot.: TNS).

Cladonia pseudostellata Asahina in J. Jap. Bot. 18: 620. 10 Nov 1942. – Lectotype (designated here): Japan, Hokkaido, Mt. Daisetsu, 1937, *Y. Asahina 37016* (TNS; isolectot.: US).

Cladonia psoromica Dey in Bryologist 76: 419. 27 Sep 1973. – Holotype: U.S.A., North Carolina, Ashe Co., Bluff Mountain, 2 mi. SSW of West Jefferson, 1972, *J. P. Dey 449* (DUKE).

Cladonia pulvinella S. Hammer in Mycotaxon 40: 192. 22 Mai 1991. – Holotype: U.S.A., California, Marin Co., Point Reyes National Seashore, Ridge Trail, 1987, *S. Hammer 2023* (FH; isot.: H).

Cladonia pulviniformis Ahti in Lichenologist 22: 263. Jul 1990. – Holotype: Venezuela, Bolívar, Distrito Piar, Macizo del Chimantá, Acopán-tepui, 1985, *T. Ahti & al. 45045* (VEN; isot.: B, COL, DUKE, H, MYF, NY, US).

Cladonia pycnoclada (Pers.) Nyl. in J. Linn. Soc., Bot., 9: 244. 1867 ["1866"]. – Basionym: *Cenomyce pycnoclada* Pers. in Gaudichaud, Voy. Uranie, Bot.: 212. 23 Feb 1828 ["1826"]. – Lectotype (Ahti, 1961): Falkland Islands, East Falklands, Port Louis, 1817-1820, *C. Gaudichaud 134* (G).

Cladonia pyxidata (L.) Hoffm., Deutschl. Fl. 2: 121. 1796. – Basionym: *Lichen pyxidatus* L., Sp. Pl.: 1151. 1 Mai 1753. – Neotype (designated here): Sweden, Uppland, Uppsala, Norbyskogen, 1819, *G. Wahlenberg s.n.* (UPS).

Cladonia pyxidata subsp. *chlorophaea* (Sommerf.) Arnold in Flora 67: 95. 11 Feb 1884. – Basionym and type: see under *Cladonia chlorophaea*.

Cladonia pyxidata subsp. *pocillum* (Ach.) Fink in Contr. U.S. Natl. Herb. 14: 125. 1 Jun 1910. – Basionym and type: see under *Cladonia pocillum*.

Cladonia pyxidata var. *chlorophaea* (Sommerf.) Flörke, De Cladon.: 70. Pasq. 1828. – Basionym and type: see under *Cladonia chlorophaea*.

Cladonia pyxidata var. *pocillum* (Ach.) Flot. in Linnaea 17: 19. 26-28 Apr 1843. – Basionym and type: see under *Cladonia pocillum*.

Cladonia ramulosa (With.) J. R. Laundon in Lichenologist 16: 225. Oct 1984. – Basionym: *Lichen ramulosus* With., Bot. Arr. Veg. Gr. Brit.: 723. 1776. – Holotype: Dillenius, Hist. Musc.: t. 15, f. 20. 1742. Typotype: England, London, Greenwich, Woolwich Heath, *Dillenius s.n.* (OXF).

Cladonia rangiferina (L.) F. H. Wigg., Prim. Fl. Holsat.: 90. 29 Mar 1780. – Basionym: *Lichen rangiferinus* L., Sp. Pl.: 1153. 1 Mai 1753. – Lectotype (Nourish & Oliver, 1974): [Sweden?], Herb. Linnaeus No. 1273.240 (LINN).

Cladonia rangiferina subsp. ***grisea*** Ahti in Ann. Bot. Soc. Zool.-Bot. Fenn. "Vanamo" 32(1): 96. 25 Sep 1961. – Holotype: Japan, Pref. Yamanashi, Lake Yamanake at foot of Mt. Fuji, 1952, *Y. Asahina s.n.* (TNS; isot.: H).

Cladonia rangiferina var. ***abbayesii*** Ahti in Ann. Bot. Soc. Zool.-Bot. Fenn. "Vanamo" 32(1): 94. 25 Sep 1961. – Holotype: Colombia, Cundinamarca, Bogotá, 1860-1870, *A. Lindig 2513* (UPS; isot.: BM, H-NYL No. 1184, M, S).

Cladonia rangiformis Hoffm., Deutschl. Fl. 2: 114. 1796. – Neotype (designated here): The Netherlands ("Holland"), *G. F. Hoffmann s.n.* (MW-Hoffm No. 8617 p.p.).

Cladonia rappii A. Evans, Trans. Connecticut Acad. Arts 38: 297. Sep 1952. – Holotype: U.S.A., Florida, Seminole Co., Sanford, 1924, *S. Rapp* in Sandstede, Cladon. Exsicc. No. 1398 p.p. (US; isot.: UPS).

Cladonia rappii var. ***exilior*** (Abbayes) Ahti in Ann. Bot. Fenn. 20: 5. 28 Apr 1983. – Basionym: *Cladonia calycantha* var. *exilior* Abbayes in Bryologist 52: 94. 13 Jul 1949. – Lectotype (designated here): Panama, Chiriquí, Chiriquí volcano, 1941, *P. F. Scholander*

s.n. (US; isolectot.: REN-Abbayes).

Cladonia ravenelii Tuck., Syn. N. Amer. Lich. 1: 254. Jan-Feb 1882. – Lectotype (designated here): U.S.A., Florida, 1877, *H. W. Ravenel s.n.* (FH-Tuck; isolectot.: BM).

Cladonia rei Schaer., Lich. Helv. Spic.: 34. 1823. – Holotype: Italy, *G. F. Re 124* (G).

Cladonia rhodoleuca Vain. in Acta Soc. Fauna Fl. Fenn. 4: 453. 3-31 Dec 1887. – Lectotype (designated here): Brazil, Minas Gerais, Caraça, 1885, *E. A. Vainio s.n.* (TUR-V No. 15172)

Cladonia rigida (Hook. f. & Taylor) Hampe in Linnaea 28: 216. Sep 1856. – Basionym: *Cenomyce rigida* Hook. f. & Taylor in London J. Bot. 3: 652. Dec 1844. – Lectotype (Galloway, 1985): New Zealand, Auckland Islands, ("Lord Auckland's group"), *J. D. Hooker 1575* (BM; isolectot.: BM, FH-Taylor, H-NYL No. 38432).

Cladonia rigida var. ***acuta*** (Taylor) A. W. Archer in Muelleria 7: 175. 30 Mar 1990. – Basionym: *Cenomyce acuta* Taylor in London J. Bot. 6: 186. 1847. – Holotype: "Islands of the Pacific" (i.e., Auckland Islands?), *J. D. Hooker s.n.* (FH-Taylor).

Cladonia robbinsii A. Evans in Trans. Connecticut Acad. Arts 35: 611. 1944. – Holotype: U.S.A., Connecticut, Lyme, 1931, *A. Evans 2625* (US).

Cladonia robusta Ahti in Ann. Bot. Fenn. 23: 213. 26 Nov 1986. – Holotype: Puerto Rico, Manatí, Barrio Tierras Nuevas Saliente, S of Laguna Tortuguero, 1984, *G. R. Proctor 40527* (US; isot.: H, SJ).

Cladonia rotundata Ahti in Ann. Bot. Soc. Zool.-Bot. Fenn. "Vanamo" 32(1): 29. 25 Sep 1961. – Holotype: Peru, San Martín, Roque, La Campana, 1925, *D. Melin 7* (UPS; isot.: S).

Cladonia rugulosa Ahti in Ann. Bot. Fenn. 23: 206. 26 Nov 1986. – Holotype: Venezuela, Mérida, Morro Negro, Pico de Horma, SE of Mesa Quintero, 1980, *M. López-Figueiras & H. Rodríguez 23086* (MERF; isot.: H, US, VEN).

Cladonia salmonea Stenroos in Ann. Bot. Fenn. 26: 255. 31 Aug 1989. – Holotype: Brazil, Minas Gerais, Congonhas, 1978, *K. Kalb & G. Plöbst* in Kalb, Lich. Neotr. No. 13 (H; isot.: BM, Herb. Kalb, UPS, US).

Cladonia salzmannii Nyl., Syn. Meth. Lich. 1: 214. Apr 1860. – Lectotype (designated here): Brazil, Bahia, *J. S. Blanchet s.n.* (PC-Hue; isolectot.: BM, ?G, H-NYL No. 37599, M, PC, PC-Lenorm, ?UPS).

Cladonia sandstedei Abbayes in J. Bot. 76: 349. Dec 1938. – Lectotype (Ahti, 1984): Jamaica, St. Andrew, Cinchona, 1896, *W. Harris s.n.* (BM; isolectot.: UCWI).

Cladonia santensis Tuck. in Amer. J. Sci. Arts, ser. 2, 25: 427. Mai 1858. – Holotype: U.S.A., South Carolina, Santee Canal, *H. W. Ravenel 77* (FH-Tuck; isot.: BM, ?H-NYL No. 38415, ?TUR-V No. 15152, ?US).

Cladonia sarmentosa (Hook. f. & Taylor) C. W. Dodge in Brit. Austral. New Zealand Antarct. Res. Exped. 1929-1931 Rep., Ser. B, 7: 129. 1948. – Basionym: *Cenomyce sarmentosa* Hook. f. & Taylor in London J. Bot. 3: 651. Dec 1844. – Lectotype (Ahti & al., 1990): New Zealand, Auckland Islands ("Lord Auckland's group"), 1844, *J. D. Hooker 1569* (BM; isolectot.: FH-Taylor).

Cladonia scabriuscula (Delise) Nyl. in Compt. Rend. Hebd. Séances Acad. Sci. 83: 88. Jul 1876. – Basionym: *Cenomyce scabriuscula* Delise in Duby, Bot. Gall.: 623. Mai 1830. – Type not designated.

Cladonia scholanderi Abbayes in Bryologist 52: 92. 13 Jul 1949. – Lectotype (designated here): Panama, Chiriquí, Volcán Chiriquí, 1941, *P. F. Scholander s.n.* (US).

Cladonia secundana Nyl. in Flora 57: 71. 11 Feb 1874. – Lectotype (Stenroos, 1989a): Venezuela, Amazonas, Depto. Río Negro, San Carlos de Río Negro, 1849-1855, *R. Spruce* in Spruce, Lich. Amaz. And. No. 35 (H-NYL No. 37826; isolectot.: BM, G).

Cladonia shikokiana Asahina in J. Jap. Bot. 32: 361. Dec 1957. – Holotype: Japan, Shikoku, Prov.

Awa, Naka-gun, between Sawa-dani-mura and Iwakura, 1956, *M. Togashi s.n.* (TNS; isot.: US).

Cladonia siamea Abbayes in Kew Bull. 1956: 262. 18 Oct 1956. – Holotype: Thailand, Phu Krading, Loie, 1954, Royal Forest Department's Collection No. 10791 (BM; isot.: REN-Abbayes).

Cladonia signata (Eschw.) Vain. in Meddeland. Soc. Fauna Fl. Fenn. 14: 32. 13 Mai-Sep 1886. – Basionym: *Cladonia rangiferina* var. *signata* Eschw. in Martius, Fl. Bras. Enum. Pl. 1(1): 275. Mai-Dec 1833. – Neotype (designated here): Brazil, Amazonas, Rio Cuieiras, 50 km upstream, 1974, *J. C. Ongley & J. F. Ramos P21767* (INPA; isoneot.: H, M, NY).

Cladonia simulata Robbins in Rhodora 31: 105. 5 Jun 1929. – Lectotype (designated here): U.S.A., Massachusetts, Plymouth Co.: Wareham, 1924, *C. A. Robbins s.n.* (FH).

Cladonia skottsbergii H. Magn. in Ark. Bot. 30B(3): 2. 25 Feb 1942. – Holotype: U.S.A., Hawaii, Molokai, Pepeopae Bog, 1938, *L. M. Cranwell & C. Skottsberg 5781* (S; isot.: GB).

Cladonia sobolescens Nyl. ex Vain. in Acta Soc. Fauna Fl. Fenn. 53(1): 80. 4 Nov - 2 Dec 1922. – Lectotype (Ahti, 1983): U.S.A., Tennessee, *W. W. Calkins 77* (H-NYL No. 38693).

Cladonia solida Vain. in Acta Soc. Fauna Fl. Fenn. 7(1): 246. 1-22

Nov 1890. – Lectotype (designated here): Brazil, Minas Gerais, Antônio Carlos ("Sitio"), 1885, *E. A. Vainio s.n.* (TUR-V No. 17216).

Cladonia solitaria H. Magn. in Ark. Bot. 30B(3): 3. 25 Feb 1942. – Holotype: U.S.A., Hawaii, Maui, Puu Kukui, 1938, *L. M. Cranwell & C. Skottsberg L5762-4* (S; isot.: UPS).

Cladonia southlandica W. Martin in Trans. Roy. Soc. New Zealand, Bot. 2: 42. 30 Nov 1962. – Holotype: New Zealand, Otago, Southland, near Awarua Bay, 1961, *W. Martin 8249* (CHR).

Cladonia sphacelata Vain. in Acta Soc. Fauna Fl. Fenn. 4: 456. 3-31 Dec 1887. – Lectotype (designated here): Brazil, Minas Gerais, Caraça, 1885, *E. A. Vainio s.n.* (TUR-V No. 15194).

Cladonia spiculata (Ach.) Ahti in Ann. Bot. Fenn. 23: 207. 26 Nov 1986. – Basionym: *Lichen spiculatus* Ach., Lichenogr. Suec. Prodr.: 206. 1799 ["1798"]. – Lectotype (Ahti, 1983): Jamaica, *O. P. Swartz s.n.* (H-ACH No. 1625; isolectot.: S-Sw, TUR-V No. 13595, UPS-Ach No. 916).

Cladonia spinea Ahti in Ann. Bot. Fenn. 23: 215. 26 Nov 1986. – Holotype: Venezuela, Amazonas, Depto. Atabapo, Cano Cañame, 1979, *S. S. Tillett & al. 795-130* (H; isot.: MYF, US, VEN).

Cladonia sprucei Ahti in Ann. Bot. Soc. Zool.-Bot. Fenn. "Vanamo" 32(1): 100. 25 Sep 1961. – Holotype: Brazil, Amazonas, Umirisál near Manaus, 1849-1855, *R. Spruce* in Spruce, Lich. Amaz. And. No. 17 (H-NYL No. 37625; isot.: BM, G, NY).

Cladonia squamosa Hoffm., Deutschl. Fl. 2: 125. 1796. – Neotype (designated here): Italy, Trentino-Alto Adige ("Tirolia orientalis"), Prov. Bolzano, Val di Pusteria, Casteldarre (Ehrenburg), *E. Kernstock* in Kerner, Fl. Exsicc. Austro-Hung. No. 3525 (H).

Cladonia squamosa var. ***subsquamosa*** (Nyl. ex Leight.) Vain. in Meddeland. Soc. Fauna Fl. Fenn. 6: 113. 29 Oct - 9 Nov 1881. – Basionym: *Cladonia delicata* var. *subsquamosa* Nyl. ex Leight., Lich. Fl. Gr. Brit.: 59. Sep-Oct 1871. – Lectotype (designated here): England, Leicestershire, *W. A. Leighton* in Leighton, Lich. Brit. Exsicc. No. 405 (H; isolectot.: BM, UPS).

Cladonia staufferi Abbayes in Bot. Jahrb. Syst. 86: 242. 1 Feb 1967 [& in Rev. Bryol. Lichénol. 34: 821. Jan-Jun 1967 ["1966"]]. – Holotype: Australia, Victoria, Australian Alps, Mt. Baw-Baw, 1963, *H. U. Stauffer 5458* (REN-Abbayes; isot.: BERN-Frey No. 28098).

Cladonia stellaris (Opiz) Pouzar & Vězda in Preslia 43: 196. 2 Aug 1971. – Basionym: *Cenomyce stellaris* Opiz, Böh. Phan. Crypt. Gew.: 141. Feb-Nov 1823 [and in Ponfinkl, Vollst. Umriss Topogr. Königr. Böhmen: 493. 1823]. –

Lectotype (Pouzar & Vězda, 1971): Dillenius, Hist. Musc.: t. 16, f. 29F. 1742.

Cladonia stereoclada Abbayes in Portugaliae Acta Biol., Sér. B, Sist., 1: 245. 1947 ["1946"]. – Type not designated.

Cladonia steyermarkii Ahti in Ann. Bot. Fenn. 23: 217. 26 Nov 1986. – Holotype: Venezuela, Bolívar, Distrito Piar, Macizo del Chimantá, Acopán-tepui, 1985, *T. Ahti & al. 45148* (VEN; isot.: B, COL, H, NY, TNS, US).

Cladonia strepsilis (Ach.) Grognot, Pl. Crypt. Saône-et-Loire: 85. 1863. – Basionym: *Baeomyces strepsilis* Ach., Methodus, Suppl.: 52. Mai-Dec 1803. – Lectotype (designated here): Sweden (H-ACH No. 1723A; isolectot.: ?UPS).

Cladonia stricta (Nyl.) Nyl. in Flora 52: 294. 1 Jul 1869. – Basionym: *Cladonia degenerans* var. *stricta* Nyl. in Middendorff, Reise Sibir. 4, Anh. 6(2): 4. 1867. – Lectotype (Ahti, 1978): Russia, Siberia, Taimyr Peninsula, 1843, *A. von Middendorff s.n.* (H-NYL No. 38841).

Cladonia stricta var. ***uliginosa*** Ahti in Ann. Bot. Fenn. 15: 11. 8 Mai 1978. – Holotype: Russia, Murmansk Region, Salla, Korja, at base of Mt. Nurmitunturi, 1937, *M. Laurila* in Räsänen, Lich. Fenn. Exsicc. No. 475 (H; isot.: BM, H, UPS).

Cladonia stygia (Fr.) Ruoss in Bot. Helv. 95: 241. 30 Dec 1985. – Basionym: *Cladonia rangiferina*

f. *stygia* Fr., Sched. Crit. Lich. Suec. 8-9: 22. 19 Dec 1826. – Neotype (Ahti & Hyvönen, 1985): Sweden, Södermanland, St. Malm, Sörgölet, 1918, *G. O. A. Malme* in Malme, Lich. Suec. Exsicc. No. 726 (H; isoneot.: UPS).

Cladonia subantarctica Filson & A. W. Archer in Muelleria 6: 230. 14 Mai 1986. – Holotype: Australia, Macquarie Island, Major Lake, 1964, *R. Filson & R. Peterson 6158* (MEL No. 20294).

Cladonia subcariosa Nyl. in Flora 59: 560. 11 Dec 1876. – Holotype: Cuba, Lalnar-Paneipe, *C. Wright* in Wright, Lich. Ins. Cubae, ser. 2, No. 92 (H-NYL No. 38698; isot.: FH-Tuck, G).

Cladonia subcervicornis (Vain.) Kernst., [Eur. Cladon.] in Jahresber. Staats-Oberrealschule Klagenfurt 43: 15, 32. Mai-Jul 1900. – Basionym: *Cladonia verticillata* var. *subcervicornis* Vain., Acta Soc. Fauna Fl. Fenn. 10: 197. Dec 1894. – Type not designated.

Cladonia subchordalis A. Evans in Rev. Bryol. Lichénol. 24: 133. Jan-Aug 1955. – Holotype: Argentina, Chubut, Los Alerces National Park, Lago Menéndez, Torrecillos stream, 1950, *I. M. Lamb 5917* (US; isot.: CANL, UPS).

Cladonia subconistea Asahina in J. Jap. Bot. 17: 433. 15 Aug 1941. – Holotype: Japan, Honshu, Prov. Shinano, Mt. Yatugadake, *Y. Asahina 704* (TNS).

Cladonia subdecaryana Yoshim. in J. Hattori Bot. Lab. 31: 201. 19 Nov 1968. – Holotype: Japan, Shikoku, Prov. Iyo, Mt. Higashi-Akaishi, 1956, *I. Yoshimura 2064* (TNS; isot.: NICH, TENN).

Cladonia subdelicatula Asahina, J. Jap. Bot. 38: 1. 1963. – Holotype: Brazil, Rio Grande do Sul, Mun. Venancio Aires, Faxinal Tamanco, 1909, *C. Jürgens 5735 (135)* (TUR-V No. 15098a; isot.: FH, US).

Cladonia subdigitata Nyl. in Compt. Rend. Hebd. Séances Acad. Sci. 83: 88. Jul 1876. – Lectotype (Ahti in Archer, 1988): New Zealand, Campbell Island, 1874, *M. Filhol s.n.* (H-NYL No. 37858; isolectot.: BM, H-NYL, MEL, PC, TUR-V No. 14155 & 14157, UPS).

Cladonia subfurcata (Nyl.) Arnold in Rehm, Cladon. Exsicc. No. 263 [label]. 1885. – Basionym: *Cladonia degenerans* f. *subfurcata* Nyl. in Not. Sällsk. Fauna Fl. Fenn. Förh. 13: 320. 2-13 Mai 1874. – Lectotype (Ahti, 1967): Finland, Kittilä Lapland, Muonio, 1867, *J. P. Norrlin s.n.* (H-NYL No. 38313; isolectot.: H).

Cladonia sublacunosa Vain. in Acta Soc. Fauna Fl. Fenn. 4: 278. 3-31 Dec 1887. – Lectotype (Ahti, 1973): Austria, Tyrol, Längental near Kühtai ("Kühthei"), 1884, *F. Arnold* in Rehm, Cladon. Exsicc. No. 244 (TUR-V No. 13551; isolectot.: BM, H, H-NYL, S, UPS).

Cladonia submedusina Müller Arg. in Flora 74: 110. 16 Jan 1891.

– Holotype: Brazil, Amazonas, Umirisál near Manaus ("Barras"), 1849-1855, *R. Spruce* in Spruce, Lich. Amaz. And. No. 21 (G; isot.: BM, G, H-NYL No. 37610, NY, PC).

Cladonia subminiata Stenroos in Ann. Bot. Fenn. 26: 256. 31 Aug 1989. – Holotype: Brazil, Paraíba, Mun. Santa Rita, c. 15 km NW of Santa Rita, 1987, *T. Ahti & L. Xavier Filho 45642* (H; isot.: JPB).

Cladonia submitis A. Evans in Rhodora 45: 435. 13 Nov 1943. – Holotype: U.S.A., Massachusetts, Plymouth Co., Wareham, 1924, *C. A. Robbins* in Sandstede, Cladon. Exsicc. No. 1391 (US; isot.: FH, H, TUR-V, UPS).

Cladonia submultiformis Asahina in J. Jap. Bot. 18: 624. 10 Nov 1942. – Lectotype (designated here): China, Taiwan, Kaohsiung ("Takao"), Mt. Dasulinshan ("Daijurin-zan"), 1941, *S. Inumaru 5970* (TNS; isolectot.: REN-Abbayes).

Cladonia subpityrea Sandst. in Keissler, Kryptog. Exsicc. Mus. Vindob. No. 3056. 1928 [and in Ann. Naturhist. Mus. Wien 41: 61. Dec 1928]. – Lectotype (Nuno, 1976): Philippines, Luzon, Manila, Dumulmog, *E. D. Merrill* in Keissler, Kryptog. Exsicc. Mus. Vindob. No. 3056 (W; isolectot.: BP, GB, H).

Cladonia subpungens Abbayes in Rev. Bryol. Lichénol. 33: 235. 1964. – Holotype: South Africa, Cape Province, District Wellington, Bains Kloof, 1953, *O. Alm-*

born 5067 (LD; isot.: H, REN-Abbayes, UPS).

Cladonia subradiata (Vain.) Sandst. in Abh. Naturwiss. Vereine Bremen 25: 230. 1922. – Basionym: *Cladonia fimbriata* subvar. *subradiata* Vain. in Acta Soc. Fauna Fl. Fenn. 10: 338. Dec 1894. – Lectotype (designated here): Brazil, Minas Gerais, Caraça, 1885, *E. A. Vainio s.n.* (TUR-V No. 19517).

Cladonia subrangiformis Sandst. in Abh. Naturwiss. Vereine Bremen 25: 165. 1922. – Lectotype (designated here): Germany, Baden-Württemberg, Wertheim, Kahlberg, 1921, *A. Kneucker* in Sandstede, Cladon. Exsicc. No. 784 (H; isolectot.: FH, TUR-V No. 15741, UPS).

Cladonia subreticulata Ahti in Ann. Bot. Fenn. 10: 168. 27 Nov 1973. – Holotype: Brazil, Minas Gerais, Caraça, 1885, *E. A. Vainio s.n.* and in Sandstede, Cladon. Exsicc. No. 1194 (TUR-V No. 13546; isot.: FH, BM, H, TUR-V, UPS, US).

Cladonia subsetacea Robbins ex A. Evans in Bryologist 50: 30. 15 Apr 1947. – Holotype: U.S.A., North Carolina, New Hanover Co.: Wrightsville, 1928, *A. Evans s.n.* (US).

Cladonia subsquamosa Kremp. in Vidensk. Meddel. Naturhist. Foren. Kjøbenhavn 5: 366. 10 Apr 1874. – Lectotype (designated here): Brazil, Minas Gerais-Rio de Janeiro, "Serra d'Estrella &

Petrópolis", *E. Warming 233* (C; isolectot.: G, M, TUR-V, UPS).

Cladonia substellata Vain. in Acta Soc. Fauna Fl. Fenn. 4: 271. 3-31 Dec 1887. – Lectotype (Ahti, 1973): Brazil, Minas Gerais, Caraça, 1885, *E. A. Vainio s.n.* (TUR-V No. 13632; isolectot.: H-NYL).

Cladonia substrepsilis Sandst. in Bot. Mag. (Tokyo) 41: 339. 1927. – Holotype: Japan, Prov. Nagano, Togakushi, 1898, *R. P. Faurie 803* (W).

Cladonia subsubulata Nyl. in Compt. Rend. Hebd. Séances Acad. Sci. 83: 88. Jul 1876. – Lectotype (designated here): New Zealand, Campbell Island, 1874, *M. Filhol s.n.* (H-NYL No. 39393).

Cladonia subtenuis (Abbayes) Mattick in Repert. Spec. Nov. Regni Veg. 49: 165. 20 Dec 1940. – Basionym: *Cladonia tenuis* subsp. *subtenuis* Abbayes in Bull. Soc. Sci. Bretagne 16, Fasc. Hors Sér. 2: 106, 108. Jul-Dec 1939. – Lectotype (Ahti, 1961): U.S.A., Florida, Seminole Co., near Sanford, *S. Rapp* in Keissler, Kryptog. Exsicc. Mus. Vindob. No. 3066 (H; isolectot.: UPS).

Cladonia subturgida Samp. in Ann. Sci. Acad. Polytecn. Porto 13: 38. Jul 1918. – Type not designated.

Cladonia subulata (L.) F. H. Wigg., Prim. Fl. Holsat.: 90. 29 Mar 1780. – Basionym: *Lichen subulatus* L., Sp. Pl.: 1153. 1 Mai 1753. – Lectotype (Laundon, 1984): [Sweden], Herb. Linnaeus No. 1273.249 (LINN).

Cladonia sufflata Ahti in Lichenologist 22: 265. Jul 1990. – Holotype: Venezuela, Bolívar, Distrito Piar, Macizo del Chimantá, Acopán-tepui, 1985, *T. Ahti & al. 45078* (VEN; isot.: B, DUKE, H, NY, US).

Cladonia sulcata A. W. Archer in Muelleria 5: 115. 22 Mar 1982. – Holotype: Australia, Victoria, 8 km E of Tawonga, Trapper's Creek Road, 1979, *A. W. Archer 803* (MEL No. 1031486; isot.: COLO, H).

Cladonia sulcata var. ***striata*** A. W. Archer in Muelleria 6: 386. 1 Apr 1987. – Holotype: Australia, New South Wales, near First Rocks, Mona Vale Road, 18, km NNW of Sydney, 1984, *A. W. Archer 1667* (MEL No. 1047761; isot.: CBG, H, NSW).

Cladonia sulcata var. ***wilsonii*** (A. W. Archer) A. W. Archer in New Zealand J. Bot. 24: 583. 13 Apr 1987. – Basionym: *Cladonia wilsonii* A. W. Archer in Muelleria 5: 274. 5 Apr 1984. – Holotype: Australia, Australian Capital Territory, 35 km SSW of Canberra, Corrin Dam Road, near Kangaroo Creek, 1982, *A. W. Archer 1315c* (MEL No. 1036222; isot.: NSW).

Cladonia sulphurina (Michx.) Fr., Lichenogr. Eur. Reform.: 237. Jun-Jul 1831. – Basionym: *Scyphophorus sulphurinus* Michx., Fl. Bor.-Amer. 2: 328. 19 Mar 1803. – Lectotype (Ahti, 1978): Canada(?) ("America Septentrionalis"), *A. Michaux s.n.* (PC-Hue; isolectot.: PC).

Cladonia symphoriza Nyl. in Ann. Sci. Nat., Bot., ser. 5, 7: 303. Mai 1867. – Lectotype (Ahti & Stenroos, 1986): Colombia, Boyacá, Muzo, 1863, *A. Lindig 2553* (H-NYL No. 1162; isolectot.: BM, FH, G, H-NYL No. 1162, M, PC, UPS).

Cladonia symphycarpa (Flörke) Fr., Sched. Crit. Lich. Suec. 8-9: 20. 19 Dec 1826. – Basionym: *Capitularia symphycarpa* Flörke in Beitr. Naturk. 2: 281. 19 Sep 1810. – Neotype (designated here): Germany, Thuringia, Nordhausen, Alter Stolberg, 1920, *H. Sandstede & K. Wein* in Sandstede, Cladon. Exsicc. No. 689 (UPS; isoneot.: H).

Cladonia tapperi Ahti & Krog in Ann. Bot. Fenn. 24: 90. 30 Jun 1987. – Holotype: Ethiopia, Bale Province, pass between Adaba and Goba, 1972, *H. Krog E 22/97* (O).

Cladonia tasmanica Ahti in Ann. Bot. Soc. Zool.-Bot. Fenn. "Vanamo" 32(1): 30. 25 Sep 1961. – Holotype: Australia, Tasmania, *C. Stuart ("V. D. Laurel") s.n.* (BM).

Cladonia tenuicaulis Nuno in J. Jap. Bot. 47: 166. Jun 1972. – Holotype: Malaysia, Pahang, Cameron Highlands, Gunong Bringchang, 1965, *H. Inoue 12221* (TNS).

Cladonia tenuis (Flörke) Harm., Lich. France: 228. Dec 1907 - Feb 1908. – Basionym and type: see under *Cladonia ciliata* var. *tenuis.*

Cladonia terrae-novae Ahti in Arch. Soc. Zool.-Bot. Fenn. "Vanamo" 14: 131. 5 Sep 1960. – Holotype: Canada, Newfoundland, Placentia West District, 4 mi. NW of Long Pond, 1956, *T. Ahti 108* (H; isot.: BM, CANL, MSC, NFLD, TNS, UPS, US, WIS).

Cladonia tessellata Ahti & Kashiw. in Inoue, Stud. Cryptog. S. Chile: 145. 30 Aug 1984. – Holotype: Chile, Llanquihue, Machihue, Río Llicó, 1981, *H. Kashiwadani 17952* (TNS; isot.: H, SGO).

Cladonia testaceopallens Vain. in Acta Soc. Fauna Fl. Fenn. 10: 26. Dec 1894. – Holotype: Brazil. Minas Gerais, Caraça, 1885, *E. A. Vainio s.n.* (TUR-V No. 17228).

Cladonia thiersii S. Hammer in Mycotaxon 34: 115. 20 Jan 1989. – Holotype: U.S.A., California, Marin Co.: Point Reyes Mational Seashore, Kehoe Beach, 1988, *S. Hammer 2286* (SFSU; isot.: FH).

Cladonia thomsonii Ahti in Bryologist 81: 334. 18 Jul 1978. – Holotype: U.S.A., Alaska, Chugach Mts., Thompson Pass on Richardson Highway, 1967, *T. Ahti & J. W. Thomson 24604* (H; isot.: CANL, WIS).

Cladonia tixieri Abbayes in Rev. Bryol. Lichénol. 32: 218. Jan-Feb 1964. – Holotype: Vietnam, Dalat region, Manline, 1959, *P. Tixier s.n.* (REN-Abbayes; isot.: TUR).

Cladonia transcendens (Vain.) Vain. in Nouv. Arch. Mus. Hist. Nat., ser. 3, 10: 262. 1898. – Basionym: *Cladonia corallifera* var. *transcendens* Vain. in Acta Soc. Fauna Fl. Fenn. 4: 179. 3-31 Dec 1887. – Neotype (designated here): Canada, British Columbia, Queen Charlotte Islands, Graham Island, McClinton Bay, 1967, *I. M. Brodo & al. 13003* (CANL; isoneot.: H).

Cladonia transindica Ahti in Ann. Bot. Soc. Zool.-Bot. Fenn. "Vanamo" 32(1): 72. 25 Sep 1961. – Holotype: Vietnam, Dalat ("Dalut"), Chu Yang Sing, 1954, *M. Schmid s.n.* (REN-Abbayes).

Cladonia turgida Hoffm., Deutschl. Fl. 2: 124. 1796. – Type not designated.

Cladonia turgidior (Nyl.) Ahti in Lichenologist 9: 14. Apr 1977. – Basionym: *Cladonia gorgonea* [unranked] *turgidior* Nyl., Syn. Meth. Lich. 1: 213. Apr 1860. – Lectotype (Ahti, 1977): Brazil, Rio de Janeiro, 1816-1821, *A. Saint-Hilaire s.n.* (PC; isolectot.: H-NYL No. 37621, PC, TUR-V No. 15039).

Cladonia umbellata Ahti & Krog in Ann. Bot. Fenn. 24: 90. 30 Jun 1987. – Holotype: Tanzania, Southern Highlands Prov., Njombe District, between Kikondo and Kitulo, 1978, *Å. E. Dahl s.n.* (O).

Cladonia umbricola Tønsberg & Ahti in Norweg. J. Bot. 27: 307. Dec 1980. – Holotype: Norway, Sør-Trøndelag, Klaebu, Ramgåa SW of Selbusjøen, 1976, *R. Hjelmstad s.n.* (TRH; isot.: BM, H, O, UPS).

Cladonia uncialis (L.) F. H. Wigg., Prim. Fl. Holsat.: 90. 29 Mar 1780. – Basionym: *Lichen uncialis* L., Sp. Pl.: 1153. 1 Mai 1753. – Neotype (designated here): Sweden, Dalarna, Stora Kopparberg, Rotneby ("Rottneby prope urbem Fahlun Dalekarliae"), *C. Stenhammar* in Stenhammar, Lich. Suec. Exsicc., ed. 2, No. 210 (UPS; isoneot.: H).

Cladonia uncialis subsp. ***biuncialis*** (Hoffm.) M. Choisy in Bull. Mens. Soc. Linn. Lyon 20: 9. 1951. – Basionym: *Cladonia biuncialis* Hoffm., Deutschl. Fl. 2: 116. 1796. – Neotype (Ahti, 1978): [Germany?], *G. F. Hoffmann s.n.* (MW-Hoffm No. 8614) [earlier designated as "lectotype"].

Cladonia usambarensis Ahti & Krog in Ann. Bot. Fenn. 24: 92. 30 Jun 1987. – Holotype: Tanzania, Tanga Prov., Lushoto District, West Usambara Mountains, Balangai West Forest Reserve, E side of Kilimandege summit, 1984, *A. Borhidi & M. Hedrén 8414/P* (VBI; isot.: O).

Cladonia ustulata (Hook. f. & Taylor) Leight. in Ann. Mag. Nat. Hist., ser. 3, 19: 109. 1867. – Basionym: *Cenomyce ustulata* Hook. f. & Taylor in London J. Bot. 3: 652. Dec 1844. – Lectotype (Ahti & al., 1990): Falkland Islands, Uranie Bay, 1844, *J. D. Hooker s.n.* (BM; isolectot.: FH-Taylor, PC-Mont).

Cladonia valida (Abbayes) Stenroos in Ann. Bot. Fenn. 28: 108. 3

Jul 1991. – Basionym: *Cladonia diplotypa* f. *valida* Abbayes in Rev. Bryol. Lichénol. 16: 86. 1947. – Holotype: Madagascar, Fianarantsoa, Bassin de l'Itomampy, Mont Papanga near Béfotaka, 1928, *H. Humbert s.n.* (PC; isot.: REN-Abbayes).

Cladonia vareschii Ahti in Ann. Bot. Fenn. 23: 218. 26 Nov 1986. – Holotype: Venezuela, Bolívar, Distrito Piar, Auyántepui, El Peñon, 1975, *V. Vareschi 8704* (VEN; isot.: H).

Cladonia varians Vain. ex Ahti in Ann. Bot. Fenn. 24: 93. 30 Jun 1987. – Holotype: Réunion, 1890, *Frère Rodríguez s.n.* (TUR-V No. 15077; isot.: PC).

Cladonia variegata Ahti in Lichenologist 22: 266. Jul 1990. – Holotype: Venezuela, Bolívar, Distrito Piar, Macizo del Chimantá, Churí-tepui, 1985, *T. Ahti & al. 44912* (VEN; isot.: B, BM, COL, DUKE, H, MYF, NY, US).

Cladonia verruculosa (Vain.) Ahti in Bryologist 81: 336. 18 Jul 1978. – Basionym: *Cladonia pityrea* var. *verruculosa* Vain. in Acta Soc. Fauna Fl. Fenn. 10: 355. Dec 1894. – Lectotype (designated here): Canada, British Columbia, Vancouver Island, 1858-1859, *D. Lyall s.n.* (PC; isolectot.: FH-Tuck).

Cladonia verticillaris (Raddi) Fr., Lichenogr. Eur. Reform.: 465. Jun-Jul 1831. – Basionym: *Cenomyce verticillaris* Raddi, Alc. Sp. Rett. Piant. Bras.: 34. 1820. – Type not designated.

Cladonia verticillata (Hoffm.) Schaer., Lich. Helv. Spic.: 31. 1823. – Basionym: *Cladonia pyxidata* [unranked] *verticillata* Hoffm., Deutschl. Fl. 2: 122. 1796. – Neotype (designated here): [Germany?], *G. F. Hoffmann s.n.* (MW-Hoffm No. 8642).

Cladonia vulcani Savicz in Izv. Imp. Bot. Sada Petra Velikago 14: 124. 1914. – Holotype: Russia, Kamchatka Region, volcano Uson, 1909, *V. P. Savicz 6413* (LE; isot.: H, UPS, US).

Cladonia vulcanica Zoll. & Moritzi in Natuur- Geneesk. Arch. Ned.-Indië 1: 396. 1844 [and in Flora 30: 317. 28 Mai 1847]. – Lectotype (Stenroos, 1986): Indonesia, Java, Banjoewangie, 1845, *H. Zollinger 87* (L).

Cladonia wainioi Savicz in Izv. Imp. Bot. Sada Petra Velikago 14: 125. 1914. – Holotype: Russia, Kamchatka Region, between Malka and Ganal, 1909, *V. P. Savicz 5244* (LE; isot.: FH, H, UPS, US).

Cladonia weymouthii F. Wilson ex A. W. Archer in Muelleria 6: 94. 29 Mai 1985. – Holotype: Australia, Tasmania, Huon River, 1892, *W. A. Weymouth s.n.* (MEL No. 6790; isot.: NSW).

Cladonia yunnana (Vain.) Abbayes ex J. C. Wei & Y. M. Jiang, Lich. Xizang: 84. Apr 1986 [and Stenroos in Ann. Bot. Fenn. 23: 246. 26 Nov 1986]. – Basionym: *Cladonia transcendens* var. *yunnana* Vain. in Nouv. Arch. Mus. Hist. Nat., ser. 3, 10: 262. 1898. –

Lectotype (designated here): China, Yunnan, Mt. Tsang-chan, 1883, *R. P. Delavay s.n. (137* in TUR-V) (PC-Hue; isolectot.: TUR-V No. 14169).

Cladonia zopfii Vain. in Meddeland. Soc. Fauna Fl. Fenn. 45: 4, 306. 1920. – Lectotype (designated here): Germany, Niedersachsen, Oldenburg, Zwischenahn, Ohrwege, Kehnmoor, 1886, *H. Sandstede* in Zwackh, Lich. Exsicc. No. 996 (TUR-V No. 13875; isolectot.: H-NYL No. 37662, UPS).

Gymnoderma Nyl.

Gymnoderma coccocarpum Nyl. in Flora 43: 546. 21 Sep 1860. – Lectotype (designated here): India, Sikkim, Tonglo ("Tongloo"), *J. D. Hooker 2124* (PC; isolectot.: BM, H-NYL No. 30796 & 30953, LE, US).

Gymnoderma insulare Yoshim. & Sharp in Amer. J. Bot. 55: 638. 17 Mai 1968. – Holotype: Japan, Hondo, Prov. Kii, Mt. Koya, 1957, *S. Kurokawa 57277* in Kurokawa, Lich. Rar. Crit. Exsicc. No. 8 (TNS; isot.: DUKE, H, M, NICH, UPS, US).

Gymnoderma lineare (A. Evans) Yoshim. & Sharp in Amer. J. Bot. 55: 639. 17 Mai 1968. – Basionym: *Cladonia linearis* A. Evans in Bryologist 50: 46. 15 Apr 1947. – Holotype: U.S.A., Tennessee, Sevier ("Monroe") Co., Roaring Fork, Mt. LeConte, 1933, *K. Baur & M. Fulford 39* (US).

Gymnoderma melacarpum (F. Wilson) Yoshim. in J. Jap. Bot. 48: 287. Sep 1973. – Basionym and type: see under *Neophyllis melacarpa*.

Heteromyces Müller Arg.

Heteromyces rubescens Müller Arg. in Flora 72: 505. 20 Dec 1889. – Holotype: Brazil, Rio de Janeiro, *E. Ule 38* (G).

Metus D. J. Galloway & P. James

Metus conglomeratus (F. Wilson) D. J. Galloway & P. James in Notes Roy. Bot. Gard. Edinburgh 44: 566. 10 Dec 1987. – Basionym: *Pilophoron conglomeratum* F. Wilson in Victorian Naturalist 6: 68. 1889. – Holotype: Australia, Victoria, Black Spur, *F. R. M. Wilson 70* (BM).

Metus efflorescens D. J. Galloway & P. James in Notes Roy. Bot. Gard. Edinburgh 44: 569. 10 Dec 1987. – Holotype: Chile, IX Region, Parque Nac. Conguillio, Laguna Captrén, 1986, *B. J. Coppins & al. 4001* (SGO; isot.: BM, E).

Metus pileatus (Mont.) D. J. Galloway & P. James in Notes Roy. Bot. Gard. Edinburgh 44: 571. 10 Dec 1987. – Basionym: *Cladonia pileata* Mont. in Ann. Sci. Nat., Bot., ser. 3, 18: 310. 1852. – Lectotype (designated here): Chile, (Coquimbo?), *C. Gay s.n.* (PC-Mont; isolectot.: H-NYL No. 39141, M).

Myelorrhiza Verdon & Elix

Myelorrhiza antrea Verdon & Elix in Brunonia 9: 195. 16 Mar 1987. – Holotype: Australia, Queensland, Cardwell Range, Bishop Peak, 1984, *J. A. Elix & H. Streimann 15809* (CBG).

Myelorrhiza jenjiana Verdon & Elix in Brunonia 9: 196. 16 Mar 1987. – Holotype: Australia, Queensland, Lannercost State Forest, Blue Water Creek, 1984, *J. A. Elix & H. Streimann 15496* (CBG).

Neophyllis F. Wilson

Neophyllis melacarpa (F. Wilson) F. Wilson in J. Linn. Soc., Bot., 28: 372. 31 Oct 1891. – Basionym: *Phyllis melacarpa* F. Wilson in Victorian Naturalist 6: 68. 1889. – Holotype: Australia, Victoria, Black Spur, 1888, *F. R. M. Wilson s.n.* (NSW; isot.: BM, G).

Neophyllis pachyphylla (Müller Arg.) Gotth. Schneid. in Biblioth. Lichenol. 13: 168. 1979. – Basionym: *Psora pachyphylla* Müller Arg. in Flora 70: 319. 11 Jul 1887. – Holotype: Australia, Victoria, Mt. William, *D. Sullivan 86* (G).

Pycnothelia Dufour

Pycnothelia caliginosa D. J. Galloway & P. James in Notes Roy. Bot. Gard. Edinburgh 44: 573. 10 Dec 1987. – Holotype: New Zealand, South Island, Nelson, Denniston Plateau N of Westport, 1981, *D. J. Galloway s.n.* (CHR No. 381055; isot.: BM, CHR, H).

Pycnothelia papillaria Dufour in Ann. Gén. Sci. Phys. 8: 46. Mai 1821. – Holotype: Dillenius, Hist. Musc.: t. 16, f. 28. 1742. Typotype: England, Surrey, Bagshot, *J. J. Dillenius s.n.* (OXF).

Thysanothecium Mont. & Berk.

Thysanothecium hookeri Mont. & Berk. in London J. Bot. 5: 257. 1846. – Lectotype (Galloway, 1977b): Australia, Western Australia, Swan River, *J. Drummond 69* (BM; isolectot.: E, H-NYL No. 40230, PC, UPS).

Thysanothecium scutellatum (Fr.) D. J. Galloway in Nova Hedwigia 36: 390. 31 Mar 1983 ["1982"]. – Basionym: *Cladonia scutellata* Fr. in Lehmann, Pl. Preiss. 2: 141. 26-28 Nov 1846. – Lectotype (Galloway & Bartlett, 1983): Australia, Western Australia, southwestern part, 1839, *L. Preiss 2696* (UPS; isolectot.: G, MEL No. 6750 & 6751).

References

Ahti, T. 1961. Taxonomic studies on reindeer lichens (*Cladonia,* subgenus *Cladina*). *Ann. Soc. Zool.-Bot. Fenn. "Vanamo"* 32(1): 1-160.

– 1965. Some notes on British *Cladoniae. Lichenologist* 3: 84-88.

– 1966. Correlation of the chemical and morphological characters in *Cladonia chlorophaea* and allied lichens. *Ann. Bot. Fenn.* 3: 380-390.

– 1967. Nomenclatural notes on *Cladonia delessertii* and *Cladonia alpicola. Bryologist* 70: 104-105.

– 1973. Taxonomic notes on some species of *Cladonia,* subsect. *Unciales. Ann. Bot. Fenn.* 10: 163-184.

– 1977. The *Cladonia gorgonina* group and *C. gigantea* in East Africa. *Lichenologist* 9: 1-15.

– 1978. Nomenclatural and taxonomic remarks on European species of *Cladonia. Ann. Bot. Fenn.* 15: 7-14.

– 1980. Taxonomic revision of *Cladonia gracilis* and its allies. *Ann. Bot. Fenn.* 17: 195-243.

– 1983. Taxonomic notes on some American species of the lichen genus *Cladonia. Ann. Bot. Fenn.* 20: 1-7.

– 1984. The status of *Cladina* as a genus segregated from *Cladonia. Beih. Nova Hedwigia* 79: 25-71.

– 1986. New species and nomenclatural combinations in the lichen genus *Cladonia. Ann. Bot. Fenn.* 23: 205-220.

– & Aptroot, A. 1992. Lichens of Madagascar: *Cladoniaceae. Cryptog. Bryol. Lichénol.* 13: 117-124.

– & Hyvönen, S. 1985. *Cladina stygia,* a common, overlooked species of reindeer lichen. *Ann. Bot. Fenn.* 22: 223-229.

– & Kashiwadani, H. 1984. The lichen genera *Cladia, Cladina* and *Cladonia* in southern Chile. Pp. 125-151 *in*: Inoue, H. (ed.), *Studies on cryptogams in southern Chile.* Tokyo.

– , Krog, H. & Swinscow, T. D. V. 1987. New or otherwise interesting *Cladonia* species in East Africa. *Ann. Bot. Fenn.* 24: 85-94.

– & Stenroos, S. 1986. A revision of *Cladonia* sect. *Cocciferae* in the Venezuelan Andes. *Ann. Bot. Fenn.* 23: 229-238.

– , – & Archer, A. W. 1990. Some species of *Cladonia,* published by J. D. Hooker & T. Taylor ffrom the Southern Hemisphere. *Muelleria* 7: 173-177.

Archer, A. W. 1986a. Nomenclatural notes on some Australian *Cladonia* species. *Lichenologist* 18: 241-246.

– 1986b. he chemistry and distribution of *Cladonia capitellata* (J. D. Hook. & Taylor) Church. Bab. (Lichenes) in Australia. *Proc. Linn. Soc. New South Wales* 108: 191-194.

– 1988. The lichen genus *Cladonia* section *Cocciferae* in Australia. *Proc. Linn. Soc. New South Wales* 110: 205-213.

Evans, A. W. 1944. On *Cladonia polycarpia* Merrill. *Bryologist* 47: 49-56.

Filson, R. B. 1981. A revision of the lichen genus *Cladia* Nyl. *J. Hattori Bot. Lab.* 49: 1-75.

Galloway, D. J. 1977a. Additional notes on the lichen genus *Cladia* Nyl., in New Zealand. *Nova Hedwigia* 28: 475-486.

– 1977b. The lichen genus *Thysanothecium* Mont. & Berk., an historical note. *Nova Hedwigia* 28: 499-513.

– 1985. *Flora of New Zealand lichens.* Wellington.

– & Bartlett, J. K. 1983 ["1982"]. The lichen genus *Thysanothecium* Mont. & Berk., in New Zealand. *Nova Hedwigia* 36: 381-398.

Holmgren, P. K., Holmgren, N. H. & Barnett, L. C. (ed.) 1990. Index herbariorum. Part I: the herbaria of the world. Eighth edition. *Regnum Veg.* 120.

Huovinen, K. & Ahti, T. 1986. The composition and contents of aromatic lichen substances in *Cladonia,* sect. *Unciales. Ann. Bot. Fenn.* 23: 173-188.

Jølle, O. H. 1977. Ny lav for Norge og Sverige: *Cladonia cyathomorpha. Blyttia* 35: 163-166.

Laundon, J. R. 1984. Proposal to emend *Cladonia* Hill ex Browne, 1756, nom. cons., and delete *Cladona* Adanson, 1763, nom. rej. (Ascomycetes: Lecanorales). *Taxon* 33: 109-112.

Nourish, R. & Oliver, R. W. A. 1974. Chemical studies on some lichens in the Linnaean Herbariumm and lectotypification of *Lichen rangiferinus* L. (em. Ach.). *Biol. J. Linn. Soc.* 6: 259-268.

Nuno, M. 1976. On the identity of *Cladonia subpityrea* Sandst. and *C. merrillii* Sandst. *J. Jap. Bot.* 51: 381-384.

Pouzar, Z. & Vězda, A. 1971. *Cladonia stellaris* (Opiz) Pouz. et Vězda, the correct name for *Cladonia alpestris* (L.) Rabenh. *Preslia* 43: 193-197.

Ruoss, E. 1987. Chemotaxonomische und morphologische Untersuchungen an den Rentierflechten *Cladonia arbsucula* und *C. mitis. Bot. Helv.* 97: 239-263.

– & Ahti, T. 1985. Die Rentierflechten (*Cladonia* subg. *Cladina*) im Herbarium Wallroth, Strassburg. *Nova Hedwigia* 41: 147-158.

Stenroos, S. 1986. The family *Cladoniaceae* in Melanesia. 2. *Cladonia* sect. *Cocciferae. Ann. Bot. Fenn.* 23: 239-250.

– 1988. The family *Cladoniaceae* in Melanesia. 3. *Cladonia* sections *Helopodium, Perviae,* and *Cladonia. Ann. Bot. Fenn.* 25: 117-148.

– 1989a. Taxonomic revision of the *Cladonia miniata* group. *Ann. Bot. Fenn.* 26: 237-261.

– 1989b. Taxonomy of the *Cladonia coccifera* group. 2. *Ann. Bot. Fenn.* 26: 307-317.

– 1991. Status of our species of *Cladonia* endemic to the Madagascan region. *Ann. Bot. Fenn.* 28: 107-110.

– & Ahti, T. 1991 ["1990"]. The lichen family *Cladoniaceae* in Tierra del Fuego: problematic or otherwise noteworthy taxa. *Ann. Bot. Fenn.* 27: 317-327.

– , Ferraro, L. I. & Ahti, T. 1992. Lichenes Lecanorales: *Cladoniaceae. In:* Anonymous, *Fl. Cryptog. Tierra del Fuego*, 13(7). Buenos Aires.

Suominen, J. & Ahti, T. 1966. The occurrence o the lichens *Cladonia nemoxina, C. glauca,* and *C. polycarpoides* in Finland. *Ann. Bot. Fenn.* 3: 418-423.

Tønsberg, T. 1975. *Cladonia mmetacorallifera* new to Europe. *Norweg. J. Bot.* 22: 129-132.

Verseghy, K. 1964. *Typen-Verzeichnis der Flechtensammlung in der Botanischen Abteilung des Ungarischen Naturwissenschaftlichen Museums.* Budapest.

Yoshimura, I. 1968. Lichenological notes. 1. Some species of *Cladonia* with taxonomic problems. *J. Hattori Bot. Lab.* 31: 198-204.

NAMES IN CURRENT USE IN THE
PINACEAE (GYMNOSPERMAE)
IN THE RANKS OF GENUS TO VARIETY

By Aljos Farjon

The present list covers one of the groups that had been selected to test how the NCU principle can be made to work at the infrageneric level, when names are combinations consisting of two or more elements. While it was apparent from many contributions and remarks made at the NCU symposium of 20-23 February 1991 at Kew (Hawksworth, 1991) that security in the nomenclature of plants ultimately must depend on the stability of the names of species, the prospect of granting nomenclatural protection to lists of names at this and lower ranks proved to be controversial, as is also reflected in the relevant proposals to amend the *Code* (Greuter, 1991b).

The definition of names in current use as "names that one would adopt, or that other botanists would likely adopt, when referring to a given taxon" (see e.g. Hawksworth & Greuter, 1989) poses some problems when applied to taxa at the rank of species or below, especially in groups of plants that receive taxonomic attention widely beyond the limited circles of professional taxonomists. Hunt (1991) has clearly outlined the problem for succulent plants, and much of what he writes applies to conifers as well. Some kind of screening cannot be avoided when compiling a list of names with the ultimate aim of obtaining protected status under the *Code*. There are many recent examples of names which hardly qualify for inclusion in such a list, such as new combinations (mainly reflecting changes of rank or the splitting of genera) made in an almost purely mechanical way, or newly described "species" based on very doubtful or inadequate evidence.

A name in current use in *Pinaceae* has therefore been considered to be a name (a) published in accordance with the rules of the *Code* (Greuter & al., 1988) and (b) used in a monograph, flora or checklist of relatively recent date, or in an older work still more or less frequently used and cited by botanists. Names published too recently to be cited in other relevant publications were included only when accompanied by some taxonomic or nomenclatural evidence beyond the mere diagnosis or basionym citation. When different, competing names based on the same type are in use, as is still the case in several instances, the legitimate name that has priority has been accepted, even when this implies the abandonment of a more frequently used but illegitimate alternative. In a very few cases, where valid publication of a name is in doubt but no other name exists for a generally accepted taxon, that name has nevertheless been listed. In so doing, I have anticipated that the names on this list will be granted nomenclatural protection and thus be ruled to be validly published as cited, which may avoid the need of giving the taxon a new name in the future.

Names at different ranks based on the same type have been included as often as was thought appropriate, to accommodate conflicting taxonomic views. Names in the ranks of subgenus, section, subsection, and in one case series, have been included on more restrictive criteria. No subseries are listed, and names in ranks below variety have also been excluded.

For many names of conifers, particularly the early ones, no holotype was indicated. Their lectotypification requires careful study and often complicated searches. Original collections are scattered almost world-wide, which virtually impedes their comprehensive evaluation by a single person. Some searches were carried out by the compiler of this list, but many lectotypes (or neotypes) must await designation at a later stage. The lectotypification of all Linnaean species names of conifers has been prepared recently in cooperation with C. E. Jarvis (BM), and several such lectotypes are first designated in this list. Material received on loan from TI (Hayata's names), MO (Engelmann), L and M (Siebold & Zuccarini), and several other verifications by the compiler or his correspondents, enabled the designation of further lectotypes in the frame of this list.

Technical explanations

The present list is exclusively concerned with non-fossil taxa found in the wild. Names of fossils and of taxa that originated in cultivation, e.g. many nothotaxa, are excluded.

The listed names of taxa are spelled exactly as they appear in the protologue, except in those cases in which the *Code* imposes a different spelling (termination or connecting vowel) and in a few instances of orthographic corrections that are generally accepted or have been proposed in print.

Illustrations which are part of the protologue are usually cited explicitly, especially when they may be of help in identifying a type specimen, or may themselves serve as types.

When a definite type (holotype, lectotype or neotype) is cited, whether it had been designated previously or is designated here, the country, province and place of origin, the date of collection and the collector(s) with collection number are indicated as fully as the protologue permits and as was deemed necessary for unambiguous identification. The herbarium holding the type is identified by its standard abbreviation (Holmgren & al., 1990). Herbaria holding duplicates of the type are similarly mentioned but are not listed exhaustively.

Acknowledgements

After preliminary editing, a draft of this list was distributed for comment to major botanical institutions across the world (Greuter, 1991a), and personally to several conifer specialists. All comments received were carefully considered and have resulted in numerous additions and improvements incorporated in the present list.

The following persons deserve our thanks for their contribution: Teuvo Ahti & Leena Hämet-Ahti (H), Richard K. Brummitt (K), Michael Frankis (Newcastle upon Tyne, U.K.), Werner Greuter (B), Ruurd D. Hoogland (P), Charles E. Jarvis (BM), Marie-Françoise Passini (Université Pierre et Marie Curie, Paris), Jesse P. Perry, Jr. (Hertford, NC, U.S.A.), Keith Rushforth (Fareham, Hampshire, U.K.), Peter Schmidt (Technische Universität, Dresden, Germany), Adriano Soldano (Vercelli, Italy), and Gea Zijlstra (U). The curators of many herbaria mentioned hereunder are thanked for the loan of specimens and/or access to their collections, as the case may be.

Abies Mill.

Abies subg. *Pseudotorreya* Franco in Bol. Soc. Portug. Ci. Nat. 13 (Suppl. 2): 167. 1942. – Holotype: *Abies bracteata* (D. Don) A. Poit.

Abies sect. *Amabiles* (Matzenko) Farjon & Rushforth in Notes Roy. Bot. Gard. Edinburgh 46: 69. 12 Dec 1989. – Basionym: *Abies* ser. *Amabiles* Matzenko in Novosti Sist. Vyss. Rast. 1968: 11. 18 Dec 1968. – Holotype: *Abies amabilis* Douglas ex J. Forbes.

Abies sect. *Balsameae* Engelm. in Trans. Acad. Sci. St. Louis 3: 597. 1878. – Holotype: *Abies balsamea* (L.) Mill.

Abies sect. *Bracteatae* Engelm. in Trans. Acad. Sci. St. Louis 3: 596. 1878. – Holotype: *Abies bracteata* (D. Don) A. Poit.

Abies sect. *Grandes* Engelm. in Trans. Acad. Sci. St. Louis 3: 596. 1878. – Holotype: *Abies grandis* (Douglas ex D. Don) Lindl.

Abies sect. *Momi* Franco [Abetos] in Anais Inst. Super. Agron. 17: 121. 1950. – Holotype: *Abies firma* Siebold & Zucc.

Abies sect. *Nobiles* Engelm. in Trans. Acad. Sci. St. Louis 3: 596. 1878. – Holotype: *Abies nobilis* (Douglas ex D. Don) Lindl., non A. Dietrich (*Abies procera* Rehder).

Abies sect. *Oiamel* Franco [Abetos] in Anais Inst. Super. Agron. 17: 119. 1950. – Holotype: *Abies religiosa* (Kunth) Schltdl. & Cham.

Abies sect. *Piceaster* Spach, Hist. Nat. Vég. 11: 414. 25 Dec 1841. – Holotype: *Abies pinsapo* Boiss.

Abies sect. *Pseudopicea* Hickel in Bull. Soc. Dendrol. France 2: 52, f. 2. 1906. – Holotype: *Abies webbiana* Lindl. [= *Abies spectabilis* (D. Don) Spach].

Abies subsect. *Delavayanae* Farjon & Rushforth in Notes Roy. Bot. Gard. Edinburgh 46: 70. 12 Dec 1989. – Holotype: *Abies delavayi* Franch.

Abies subsect. *Firmae* (Franco) Farjon & Rushforth in Notes Roy. Bot. Gard. Edinburgh 46: 69. 12 Dec 1989. – Basionym: *Abies* ser. *Firmae* Franco [Abetos] in Anais Inst. Super. Agron. 17: 122. 1950. – Holotype: *Abies firma* Siebold & Zucc.

Abies subsect. *Hickelianae* Farjon & Rushforth in Notes Roy. Bot. Gard. Edinburgh 46: 70. 12 Dec 1989. – Holotype: *Abies hickelii* Flous & Gaussen.

Abies subsect. *Holophyllae* Farjon & Rushforth in Notes Roy. Bot. Gard. Edinburgh 46: 71. 12 Dec 1989. – Holotype: *Abies holophylla* Maxim.

Abies subsect. *Homolepides* (Franco) Farjon & Rushforth in Notes Roy. Bot. Gard. Edinburgh 46: 69. 12 Dec 1989. – Basionym: *Abies* ser. *Homolepides* Franco [Abetos] in Anais Inst. Super. Agron. 17: 121. 1950. – Holotype: *Abies homolepis* Siebold & Zucc.

Abies subsect. *Medianae* Patschke in Bot. Jahrb. Syst. 48: 643. 14 Jan 1913. – Lectotype (Farjon & Rushforth, 1989): *Abies sachalinensis* (F. Schmidt) Mast.

Abies subsect. *Religiosae* (Matzenko) Farjon & Rushforth in Notes Roy. Bot. Gard. Edinburgh 46: 70. 12 Dec 1989. – Basionym: *Abies* ser. *Religiosae* Matzenko in Novosti Sist. Vysš. Rast. 1968: 9. 18 Dec 1968. – Holotype: *Abies religiosa* (Kunth) Schltdl. & Cham.

Abies subsect. *Squamatae* E. Murray in Kalmia 14: 8. Jan 1984. – Holotype: *Abies squamata* Mast.

Abies alba Mill., Gard. Dict., ed. 8: *Abies* No. 1. 16 Apr 1768. – Syn. subst.: *Pinus picea* L., *Sp. Pl.: 1001. 1 Mai 1753. – Lectotype (designated here by A. Farjon & C. E. Jarvis): "A 2", Herb. Clifford (HSC) 449 (p.) (BM).*

Abies alba subsp. *nebrodensis* (Lojac.) Nitz. in Lustgården 1968: 178. 1969. – Basionym and type: see under *Abies nebrodensis.*

Abies alba var. *nebrodensis* (Lojac.) Svoboda in Trudy Bot. Inst. Akad. Nauk S.S.S.R., Ser. 1, Fl. Sist. Vysš. Rast. 13: 60. 1964. – Basionym and type: see under *Abies nebrodensis.*

Abies amabilis Douglas ex J. Forbes, Pinet. Woburn.: 125, t. 44. Mai 1839. – Holotype: U.S.A., Oregon, Cascade Mts., *D. Douglas s.n.* (K).

Abies arizonica Merriam in Proc. Biol. Soc. Wash. 10: 116, f. 24-25. 1896. – Type not designated.

Abies balsamea (L.) Mill., Gard. Dict., ed. 8: *Abies* No. 3. 16 Apr 1768. – Basionym: *Pinus balsamea* L., Sp. Pl.: 1002. 1 Mai 1753. – Lectotype (designated here by A. Farjon & C. E. Jarvis): "9 (8) Balsamea", Herb. Linn. No. 1135.14 (LINN).

Abies balsamea var. *phanerolepis* Fernald in Rhodora 11: 203. Nov 1909. – Holotype: Canada, Quebec, Percé Mt., *M. L. Fernald & J. F. Collins 860* (GH).

Abies beshanzuensis M. H. Wu in Acta Phytotax. Sin. 14(2): 16, t. 1, f. 1. Nov 1976. – Holotype: China, Zhejiang, Tung Kung Range, *M. H. Wu 7511* (PE).

Abies ×borisii-regis Mattf. (pro sp.) in Notizbl. Bot. Gart. Berlin-Dahlem 9: 235. 20 Mar 1925 [& in Mitt. Deutsch. Dendrol. Ges. 35: 26. 1925]. – Lectotype (designated here): Greece, Mt. Olympos, Monastery Hagios Dionysius, 1899, *P. E. E. Sintenis 1870* (B).

Abies bornmuelleriana Mattf. in Notizbl. Bot. Gart. Berlin-Dahlem 9: 239. 20 Mar 1925 [& in Mitt. Deutsch. Dendrol. Ges. 35: 24. 1925]. – Holotype: Turkey, Bithynia, Mt. Olympus, 1899, *J. Bornmüller 5564* (B).

Abies bracteata (D. Don) A. Poit. in Rev. Hort., ser. 2, 4: 7. 1845. – Basionym: *Pinus bracteata* D. Don in Trans. Linn. Soc. London 17: 442. 21 Jun - 9 Jul 1836. – Type not designated.

Abies cephalonica Loudon in Gard. Mag. & Reg. Rural Domest. Improv. 14: 81. 1838 [& Arbor. Frutic. Brit.: 2325. 1 Jul 1838]. – Type not designated.

Abies chengii Rushforth in Notes Roy. Bot. Gard. Edinburgh 41: 333. 21 Dec 1983. – Holotype: from cultivated tree, Westonbirt Arboretum, England, *K. D. Rushforth 423* (E).

Abies chensiensis Tiegh. in Bull. Soc. Bot. France 38: 413. 1 Mar 1892. – Holotype: China, Shaanxi, Qin Ling Shan, *Abbé David 918* (P).

Abies chensiensis subsp. *salouenensis* (Bordères & Gaussen) Rushforth in Notes Roy. Bot. Gard. Edinburgh 41: 539. 11 Jul 1984. – Basionym and type: see under *Abies salouenensis*.

Abies chensiensis subsp. *yulongxueshanensis* Rushforth in Notes Roy. Bot. Gard. Edinburgh 41: 539. 11 Jul 1984. – Holotype: China, Yunnan, Lijiang Shan, *T. T. Yu 15050* (E; isot.: BM).

Abies cilicica (Antoine & Kotschy) Carrière, Traité Gén. Conif.: 229. Jan 1855. – Basionym: *Pinus (Abies) cilicica* Antoine & Kotschy in Österr. Bot. Wochenbl. 3: 409. 29 Dec 1853. – Holotype: Turkey, Bulgar Daglari, *T. Kotschy 49611* (G).

Abies cilicica subsp. *isaurica* Coode & Cullen in Notes Roy. Bot. Gard. Edinburgh 26: 167. 5 Mar 1965. – Holotype: Turkey, Antalya, Bozburun Dag, *P. H. Davis 15505* (E).

Abies coahuilensis I. M. Johnston in J. Arnold Arbor. 24: 332. 15 Jul 1943. – Holotype: Mexico, Coahuia, Charretera Canyon, *I. M. Johnston 9010* (A).

Abies colimensis Rushforth & Narave in Notes Roy. Bot. Gard. Edinburgh 46: 105. 12 Dec 1989.

– Holotype: Mexico, Jalisco, Nevada de Colima, *K. D. Rushforth 647* (E; isot.: XAL, K, E).

Abies concolor (Gordon & Glend.) Lindl. ex Hildebr. in Verh. Naturhist. Vereines Preuss. Rheinl. Westphalens 18: 261. 1861. – Basionym: *Picea concolor* Gordon & Glend., Pinetum: 155. Jun-Dec 1858. – Holotype: U.S.A., New Mexico, Santa Fe, 1847, *A. Fendler s.n.* (MO).

Abies concolor var. ***lowiana*** (Gordon) Lemmon, Cone-Bear. Trees Pacif. Slope, ed. 3: 64. Jul 1895. – Basionym: *Picea lowiana* Gordon & Glend., Pinetum, Suppl.: 53. 1862. – Type not designated.

Abies dayuanensis Q. X. Liu in Bull. Bot. Res., Harbin 8(3): 85. Jul 1988. – Holotype: China, Hunan, Ling-xian Dayuan, Xopingao, *C. M. Zhang 202* (Herb. Inst. Forest., Changsha, Hunan).

Abies delavayi Franch. in J. Bot. (Morot) 13: 255. Aug 1899. – Holotype: China, Yunnan, Tsang Shan, near Dali, *P. J. M. Delavay 1210* (P).

Abies delavayi var. ***motuoënsis*** W. C. Cheng & L. K. Fu in Acta Phytotax. Sin. 13(4): 83. Oct 1975. – Holotype: China, SE Xizang, Motuo, Chinese collector No. 1011 (PE).

Abies delavayi var. ***nukiangensis*** (W. C. Cheng & L. K. Fu) Farjon & Silba in Phytologia 68: 13. Jan 1990 [& Farjon in Regnum Veg. 121: 55. Nov 1990]. – Basionym

and type: see under *Abies nukiangensis.*

Abies densa Griff., Not. Pl. Asiat. 4: 19. 1854. – Type not designated.

Abies durangensis Martínez in Anales Inst. Biol. Univ. Nac. México 13: 621. 1942. – Holotype: Mexico, Chihuahua, Pinos Altos, Dec 1939, *J. H. Faull s.n.* (MEXU).

Abies durangensis var. ***coahuilensis*** (I. M. Johnston) Martínez, Pinác. Mexic., ed. 3: 139. 1963. – Basionym: *Abies coahuilensis* I. M. Johnston in J. Arnold Arbor. 24: 332. 15 Jul 1943. – Holotype: Mexico, Coahuia, Charretera Canyon, *I. M. Johnston 9010* (A).

Abies equi-trojani (Asch. & Sint. ex Boiss.) Mattf. in Mitt. Deutsch. Dendrol. Ges. 35: 29. 1925. – Basionym: *Abies pectinata* var. *equi-trojani* Asch. & Sint. ex Boiss., Fl. Orient. 5: 701. Apr 1884. – Type not designated.

Abies ernestii Rehder in J. Arnold Arbor. 20: 85. 26 Jan 1939. – Holotype: China, Sichuan, Ta-pao Shan, *E. H. Wilson 2090* (A; isot.: E, K).

Abies ernestii var. ***salouenensis*** (Bordères & Gaussen) W. C. Cheng & L. K. Fu in Fl. Reipubl. Pop. Sin. 7: 93. Dec 1978. – Basionym and type: see under *Abies salouenensis.*

Abies fabri (Mast.) Craib in Notes Roy. Bot. Gard. Edinburgh 11: 278. Nov 1919. – Basionym: *Keteleeria fabri* Mast. in J. Linn. Soc., Bot. 26: 555. 21 Oct 1902. –

Holotype: China, Sichuan, Emei Shan, *E. Faber 984* (K).

Abies fabri subsp. **minensis** (Bordères & Gaussen) Rushforth in Notes Roy. Bot. Gard. Edinburgh 43: 273. 5 Feb 1986. – Basionym and type: see under *Abies minensis*.

Abies fanjingshanensis W. L. Huang, Tu & Fang in Acta Phytotax. Sin. 22: 154. Apr 1984. – Holotype: China, Guizhou, Fanjing Shan, *L. Yang 83-427* (Herb. Dept. Geogr., Guiyang Teach. Coll.).

Abies fargesii Franch. in J. Bot. (Morot) 13: 256. Aug 1899. – Holotype: China, Sichuan, *P. G. Farges 908-bis* (P; isot.: K).

Abies fargesii var. **faxoniana** (Rehder & E. H. Wilson) T. S. Liu, Monogr. Gen. Abies: 151. 30 Dec 1971. – Basionym and type: see under *Abies faxoniana*.

Abies fargesii var. **sutchuenensis** Franch. in J. Bot. (Morot) 13: 256. Aug 1899. – Holotype: China, Sichuan, Cheng-kou, *P. G. Farges s.n.* (P).

Abies faxoniana Rehder & E. H. Wilson in Sargent, Pl. Wilson. 2: 42. 24 Mar 1914. – Holotype: China, Sichuan, N.E. of Songpan, *E. H. Wilson 4060* (A; isot.: BM, E, K, US).

Abies ferreana Bordères & Gaussen in Trav. Lab. Forest. Toulouse 1(4, 15): 8. 1947. – Holotype: China, Yunnan. Chungtien, Yua-Tse, 3550 m, *T. T. Yu 12326* (TLF?).

Abies firma Siebold & Zucc., Fl. Jap. 2: 15, t. 107. 1842. – Lectotype (designated here): "in Japonia", *P. F. von Siebold* comm. 1842 ex herb. Zuccarini No. 284 (M).

Abies flinckii Rushforth in Notes Roy. Bot. Gard. Edinburgh 46: 101. 12 Dec 1989. – Holotype: Mexico, Jalisco, Nevada de Colima, *K. D. Rushforth 621* (XAL; isot.: E).

Abies forrestii Coltm.-Rog. in Gard. Chron., ser. 3, 65: 150. Mar 1919 [& Craib in Notes Roy. Bot. Gard. Edinburgh 11: 279. Nov 1919]. – Holotype: China, Yunnan, Lijiang Shan, *G. Forrest 6744* (E).

Abies forrestii var. **chengii** (Rushforth) Silba in Phytologia 68: 17. Jan 1990. – Basionym and type: see under *Abies chengii*.

Abies forrestii var. **ferreana** (Bordères & Gaussen) Farjon & Silba in Phytologia 68: 17. Jan 1990 [& Farjon in Regnum Veg. 121: 59. Nov 1990]. – Basionym and type: see under *Abies ferreana*.

Abies forrestii var. **georgei** (Orr) Farjon in Regnum Veg. 121: 59. Nov 1990. – Basionym and type: see under *Abies georgei*.

Abies forrestii var. **smithii** Viguié & Gaussen in Bull. Soc. Hist. Nat. Toulouse 58: 355. 1929. – Holotype: China, Yunnan, E. slope of Lijiang Shan, *J. F. Rock 10673* (A).

Abies fraseri (Pursh) Poir. in Lamarck, Encycl., Suppl. 5: 35. 1 Nov 1817. – Basionym: *Pinus* *fraseri* Pursh, Fl. Amer. Sept.: 639. Jan 1814. – Type not designated.

Abies gamblei Hickel in Bull. Soc. Dendrol. France 70: 37. 1929. – Type not designated.

Abies georgei Orr in Notes Roy. Bot. Gard. Edinburgh 18: 1, t. 236. Apr 1933. – Holotype: China, Yunnan, Jinsha-Mekong divide, *G. Forrest 22547* (E).

Abies georgei var. ***smithii*** (Viguié & Gaussen) W. C. Cheng & L. K. Fu in Acta Phytotax. Sin. 13(4): 63. Oct 1975. – Basionym and type: see under *Abies forrestii* var. *smithii*.

Abies gracilis Kom. [Fl. Manshur. 1] in Trudy Imp. S.-Peterburgsk. Bot. Sada 20: 203. Dec 1901. – Type not designated.

Abies grandis (Douglas ex D. Don) Lindl. in Penny Cyclop. 1: 30. 1833. – Basionym: *Pinus grandis* Douglas ex D. Don in Lambert, Descr. Pinus, ed. 8°, 2: p. s.n. inter 144 et 145. 1832. – Type not designated.

Abies guatemalensis Rehder in J. Arnold Arbor. 20: 285. 19 Jul 1939. – Holotype: Guatemala, Huehuetenango, Las Cumbres del Aire, *J. H. Faull 13104* (A).

Abies guatemalensis var. ***jaliscana*** Martínez in Anales Inst. Biol. Univ. Nac. México 19: 73. 1948. – Holotype: Mexico, Jalisco, Las Mesas, near Cuale, Nov 1947, *M. Martínez 28000* (MEXU).

Abies guatemalensis var. ***tacanensis*** (Lundell) Martínez, Pinác.

Mexic., ed. 3: 129. 1963. – Basionym *Abies tacanensis* Lundell in Amer. Midl. Naturalist 23: 175. 6 Feb 1940. – Holotype: Mexico, Chiapas, Volcan de Tacaná, *E. Matuda 2367* (MEXU).

Abies hickelii Flous & Gaussen in Trav. Lab. Forest. Toulouse 1(1, 17): 1. 1932 [& in Bull. Soc. Hist. Nat. Toulouse 64: 24. 1932]. – Holotype: Mexico, Oaxaca, Sierra de San Felipe?, 1900, *C. Conzatti s.n.* (LY).

Abies hickelii var. ***oaxacana*** (Martínez) Farjon & Silba in Phytologia 68: 20. Jan 1990 [& Farjon in Regnum Veg. 121: 105. Nov 1990]. – Basionym and type: see under *Abies oaxacana*.

Abies holophylla Maxim. [Diagn. Pl. Nov. Jap. 1] in Bull. Acad. Imp. Sci. Saint-Pétersbourg 10: 487. 8 Sep 1866. – Holotype: China, "Manchuria", 8 Sep 1869, *C. J. Maximowicz s.n.* (LE).

Abies homolepis Siebold & Zucc., Fl. Jap. 2: 17, t. 108. 1842. – Lectotype (designated here): "in Japonia", *P. F. von Siebold* comm. 1842 ex herb. Zuccarini No. 290 (M).

Abies homolepis var. ***umbellata*** (Mayr) E. H. Wilson [Conif. Taxads Japan] in Publ. Arnold Arbor. 8: 58, t. 39. 30 Dec 1916. – Basionym and type: see under *Abies umbellata*.

Abies kawakamii (Hayata) T. Itô in Encycl. Jap. 2: 167. 1909. – Basionym: *Abies mariesii* var. *kawakamii* Hayata in J. Coll. Sci. Imp.

Univ. Tokyo 25(19): 223. 23 Jul 1908. – Lectotype (designated here): Taiwan, Mt. Morrison, 20 Oct 1906, *T. Kawakami & U. Mori 2369* (TI).

Abies koreana E. H. Wilson in J. Arnold Arbor. 1: 188. 23 Feb 1920. – Holotype: Korea, Cheju Island (Quelpart), *E. H. Wilson 9486* (A).

Abies lasiocarpa (Hook.) Nutt., N. Amer. Sylva 3: 138. 1849. – Basionym: *Pinus (Abies) lasiocarpa* Hook., Fl. Bor.-Amer. 2: 163. Dec 1838. – Type not designated.

Abies lasiocarpa var. *arizonica* (Merriam) Lemmon in Sierra Club Bull. 2: 167. 1898. – Basionym and type: see under *Abies arizonica*.

Abies magnifica A. Murray bis in Proc. Roy. Hort. Soc. London 3: 318, f. 25-33. 1863. – Holotype: U.S.A., California, Sierra Nevada, *J. Jeffrey* (E).

Abies magnifica var. *shastensis* Lemmon in Bienn. Rep. Calif. State Board Forest. 3: 145. 1890. – Type not designated.

Abies mariesii Mast. in Gard. Chron., ser. 2, 12: 788, f. 129. 20 Dec 1879. – Holotype: Japan, Honshu, Distr. Awomori, Mt. Nikko, 1878, *G. Maries 73* (K).

Abies marocana Trab. in Bull. Soc. Bot. France 53: 154, t. 3. Apr 1906. – Type not designated.

Abies mexicana Martínez in Anales Inst. Biol. Univ. Nac. México 13: 626. 1942. – Holotype: Mexico, Nuevo León, Sierra de Santa Catarina, 1939, *M. Martínez s.n.* (MEXU).

Abies minensis Bordères & Gaussen in Trav. Lab. Forest. Toulouse 1(4, 15): 10. 1947. – Holotype: China, Sichuan, Songpan, *T. T. Yu 2586* (TLF; isot.: E).

Abies nebrodensis (Lojac.) Mattei in Boll. Reale Orto Bot. Giardino Colon. Palermo 7: 64. 1908. – Basionym: *Abies pectinata* var. *nebrodensis* Lojac., Fl. Sicula 2(2): 401. 1904-1907. – Type not designated.

Abies nephrolepis (Trautv. ex Maxim.) Maxim. [Diagn. Pl. Nov. Jap. 1] in Bull. Acad. Imp. Sci. Saint-Pétersbourg 10: 486. 8 Sep 1866. – Basionym: *Abies sibirica* var. *nephrolepis* Trautv. ex Maxim. [Prim. Fl. Amur.] in Mém. Acad. Imp. Sci. St.-Pétersbourg Divers Savans 9: 206. 1859. – Type not designated.

Abies nordmanniana (Steven) Spach, Hist. Nat. Vég. 11: 418. 25 Dec 1841. – Basionym: *Pinus nordmanniana* Steven in Bull. Soc. Imp. Naturalistes Moscou 11: 45, t. 2. 1838. – Holotype: Caucasus, source of Kur River, 1836, *A. von Nordmann s.n.* in herb. Steven (H).

Abies nordmanniana subsp. *bornmuelleriana* (Mattf.) Coode & Cullen in Notes Roy. Bot. Gard. Edinburgh 26: 167. 5 Mar 1965. – Basionym and type: see under *Abies bornmuelleriana*.

Abies nordmanniana subsp. *equitrojani* (Asch. & Sint. ex Boiss.)

Coode & Cullen in Notes Roy. Bot. Gard. Edinburgh 26: 167. 5 Mar 1965. – Basionym and type: see under *Abies equi-trojani.*

Abies nordmanniana var. ***bornmuelleriana*** (Mattf.) Silba in Phytologia 68: 21. Jan 1990. – Basionym and type: see under *Abies bornmuelleriana.*

Abies nukiangensis W. C. Cheng & L. K. Fu in Acta Phytotax. Sin. 13(4): 83. Oct 1975. – Holotype: China, Yunnan, Nukiang River, *Feng 8025* (PE).

Abies numidica de Lannoy ex Carrière in Rev. Hort. 37: 106. 1866. – Type not designated.

Abies oaxacana Martínez in Anales Inst. Biol. Univ. Nac. México 19: 39. 1948. – Holotype: Mexico, Oaxaca, Oct 1942, *M. Martínez 27900* (MEXU).

Abies ×***phanerolepis*** (Fernald) T. S. Liu, Monogr. Gen. Abies: 316. 30 Dec 1971. – Basionym and type: see under *Abies balsamea* var. *phanerolepis.*

Abies pindrow (Royle ex D. Don) Royle, Ill. Bot. Himal. Mts. 1: t. 86. Mai 1836. – Basionym: *Pinus pindrow* Royle ex D. Don in London Edinburgh Philos. Mag. & J. Sci. 8: 255. Mar 1836. – Lectotype (D. Don in Lambert, 1837): Royle, Ill. Bot. Himal. Mts. 1: t. 86. Mai 1836.

Abies pindrow var. ***brevifolia*** Dallim. & A. B. Jacks., Handb. Conif.: 126. Sep-Dec 1923. – Type not designated.

Abies pinsapo Boiss., Notice Abies Pinsapo: 8. Feb 1838 [& in Ann. Sci. Nat., Bot., ser. 2, 9: 167-172. Mar 1838; & in Biblioth. Universelle Genève 13: 406. Apr 1838]. – Type not designated.

Abies pinsapo var. ***marocana*** (Trab.) Ceballos & Bolaño in Bol. Inst. Nac. Invest. Agron. 1(2): 18. 1928. – Basionym and type: see under *Abies marocana.*

Abies pinsapo var. ***tazaotana*** (S. Côzar ex Villar) Pourtet in Ann. Ecole Natl. Eaux 9(1): 100. 1954. – Basionym and type: see under *Abies tazaotana.*

Abies procera Rehder in Rhodora 42: 522. Dec 1940. – Syn. subst.: *Pinus nobilis* (Douglas ex D. Don) Lindl. in Penny Cyclop. 1: 30. 1833, non A. Dietrich, Fl. Berlin: 793. Apr-Dec 1824 (*Pinus nobilis* Douglas ex D. Don in Lambert, Descr. Pinus, ed. 8°, 2: p. s.n. inter 144 et 145. 1832). – Type not designated.

Abies recurvata Mast. in J. Linn. Soc., Bot. 37: 423. 1 Nov 1906. – Holotype: China, Sichuan, S of Songpan, Sep 1903, *E. H. Wilson 3021* (K; isot.: BM).

Abies recurvata var. ***ernestii*** (Rehder) C. T. Kuan, Fl. Sichuan. 2: 46. 1 Jan 1983. – Basionym and type: see under *Abies ernestii.*

Abies religiosa (Kunth) Schltdl. & Cham. in Linnaea 5: 77. Jan 1830. – Basionym: *Pinus religiosa* Kunth in Humboldt & al., Nov. Gen. Sp. Pl. 2, ed. 4°: 5. 28 Apr 1817. – Holotype: Mexico,

Guerrero, *F. W. H. A. von Humboldt & A. J. A. Bonpland s.n.* (P).

Abies sachalinensis (F. Schmidt) Mast. in Gard. Chron., ser. 2, 12: 588. 8 Nov 1879. – Basionym: *Abies veitchii* var. *sachalinensis* F. Schmidt [Reis. Amur-Lande, Bot.] in Mém. Acad. Imp. Sci. Saint Pétersbourg, ser. 7, 12(2): 175. Sep-Oct 1868. – Type not designated.

Abies sachalinensis var. *gracilis* (Kom.) Farjon in Regnum Veg. 121: 83. Nov 1990. – Basionym and type: see under *Abies gracilis*.

Abies sachalinensis var. *mayriana* Miyabe & Kudô in Trans. Sapporo Nat. Hist. Soc. 7: 131. 1919. – Type not designated.

Abies sachalinensis var. *nemorensis* Mayr, Monogr. Abietin. Japan. Reich.: 42, t. 3, f. 6. Nov-Dec 1890. – Type not designated.

Abies salouenensis Bordères & Gaussen in Trav. Lab. Forest. Toulouse 1(4, 15): 4. 1947. – Holotype: China, Yunnan, near Atuntze, *T. T. Yu 7952* (TLF; isot.: A, BM, E).

Abies semenovii B. Fedtsch., Bot. Centralbl. 73: 210. 9 Feb 1898. – Holotype: Kirgizistan, Talasskij Ala Tau, Sep 1897, *V. A. Kallaur s.n.* (G).

Abies sibirica Ledeb., Fl. Altaica 4: 202. Jul-Dec 1833. – Holotype: Siberia, Altai Mts., *C. F. von Ledebour s.n.* (LE).

Abies sibirica subsp. *semenovii* (B. Fedtsch.) Farjon in Regnum Veg. 121: 81. Nov 1990. – Basionym

and type: see under *Abies semenovii*.

Abies sibirica var. *semenovii* (B. Fedtsch.) T. S. Liu, Monogr. Gen. Abies: 188. 30 Dec 1971. – Basionym and type: see under *Abies semenovii*.

Abies spectabilis (D. Don) Spach, Hist. Nat. Vég. 11: 422. 25 Dec 1841. – Basionym: *Pinus spectabilis* D. Don, Prodr. Fl. Nepal.: 55. 26 Jan - 1 Feb 1825. – Holotype: "6058", Nepal, Kumaon, *W. S. Webb* ex herb. Wallich (K).

Abies squamata Mast. in Gard. Chron., ser. 3, 39: 299, f. 121. 12 Mai 1906 [& in J. Linn. Soc., Bot. 3: 423. 1 Nov 1906]. – Holotype: China, Sichuan, W of Tachien-lu (Kangding), June 1904, *E. H. Wilson 3019* (K).

Abies tazaotana S. Côzar ex Villar, Types Sols Afrique N. 1: 80. 1947. – Type not designated.

Abies ×umbellata Mayr, Monogr. Abietin. Japan. Reich.: 34. Jan-Feb 1890 (pro sp.). – Type not designated.

Abies veitchii Lindl. in Gard. Chron. 1861: 23. 12 Jan 1861. – Type not designated.

Abies veitchii var. *sikokiana* (Nakai) Kusaka, Conif. Jap. Ill., ed. 2: 212. 1954. – Basionym: *Abies sikokiana* Nakai in Bot. Mag. (Tokyo) 42: 452. 1928. – Type not designated.

Abies vejarii Martínez, Anal. Inst. Biol. Univ. Nac. México 13: 629. 1942. – Holotype: Mexico, Tamaulipas, 20 km N. of Miquihuana, *M. Martínez 3531* (MEXU).

Abies vejarii subsp. ***mexicana*** (Martínez) Farjon in Regnum Veg. 121: 103. Nov 1990. – Basionym and type: see under *Abies mexicana.*

Abies vejarii var. ***macrocarpa*** Martínez in Anales Inst. Biol. Univ. Nac. México 19: 90. 1948. – Holotype: Mexico, Coahuila, Mesa de las Tablas, *E. E. M. Loock 123* (MEXU).

Abies vejarii var. ***mexicana*** (Martínez) T. S. Liu, Monogr. Gen. Abies: 261. 30 Dec 1971. – Basionym and type: see under *Abies mexicana.*

Abies yuanbaoshanensis Y. J. Lu & L. K. Fu in Acta Phytotax. Sin. 18: 206. Mai 1980. – Holotype: China, Guangxi, Yuanbao Shan, *Y. J. Lü 1001* (PE).

Abies ziyuanensis L. K. Fu & S. L. Mo in Acta Phytotax. Sin. 18: 208. Mai 1980. – Holotype: China, Hunan, Yinzhulao Shan, *Y. J. Lü 78001* (PE).

Cathaya Chun & Kuang

Cathaya argyrophylla Chun & Kuang in Acta Bot. Sin. 10: 246. Sep 1962 [& in Bot. Žurn. (Moscow & Leningrad) 43: 464. 14 Apr 1958, nom. inval.]. – Holotype: China, Guangxi, Lungsheng Hsien, Chinese collecting expedition Kwangfu Lingchü No. 198 (IBSC).

Cedrus Trew, *nom. cons.*

Cedrus atlantica (Endl.) G. Manetti ex Carrière, Traité Gén. Conif.: 285. Jan 1855. – Basionym: *Pinus atlantica* Endl., Syn. Conif.: 137. Mai-Jun 1847. – Type not designated.

Cedrus brevifolia (Hook. f.) A. Henry in Elwes & Henry, Trees Great Britain: 467. Apr 1908 [& Dode in Bull. Soc. Dendrol. France 1908: 59. 1908]. – Basionym and type: see under *Cedrus libani* var. *brevifolia.*

Cedrus deodara (Roxb.) G. Don in Loudon, Hort. Brit. 1: 388. 1830. – Basionym: *Pinus deodara* Roxb., Hort. Bengal.: 69. Jun-Dec 1814. – Type not designated.

Cedrus libani A. Richard in Bory, Dict. Class. Hist. Nat. 3: 299. 6 Sep 1823. Syn subst.: *Pinus cedrus* L., Sp. Pl.: 1001. 1 Mai 1753. – Lectotype (designated here by A. Farjon & C. E. Jarvis): illustration of "Cedrus" in Clusius, Exot. Libri: 162. 1605.

Cedrus libani subsp. ***atlantica*** (Endl.) Batt. & Trab., Fl. Algérie Tunisie: 397. Jan-Apr 1905. – Basionym and type: see under *Cedrus atlantica.*

Cedrus libani subsp. ***brevifolia*** (Hook. f.) Meikle, Fl. Cyprus 1: 22. 1977. – Basionym and type: see under *Cedrus libani* var. *brevifolia.*

Cedrus libani subsp. ***stenocoma*** (O. Schwarz) P. H. Davis in J. Roy. Hort. Soc. 74: 113. Mar 1949. – Basionym: *Cedrus libanitica* subsp. *stenocoma* O. Schwarz in Feddes Repert. Spec. Nov. Regni Veg. 54: 26. 30 Sep 1944. – Holotype: Turkey, Cal Dag, *O. Schwarz 461* (not traced).

Cedrus libani var. ***atlantica*** (Endl.) Hook. f. [Cedars Lebanon] in Nat. Hist. Rev., ser. 2, 2: 15. Jan 1862. – Basionym and type: see under *Cedrus atlantica.*

Cedrus libani var. ***brevifolia*** Hook. f. in J. Bot. 18: 31. Jan 1880 [& in J. Linn. Soc., Bot. 17: 518. 1880]. – Type not designated.

Hesperopeuce (Engelm.) Lemmon

Hesperopeuce mertensiana (Bong.) Rydb. in Bull. Torrey Bot. Club 39: 100. 13 Apr 1912. – Basionym and type: see under *Tsuga mertensiana.*

Keteleeria Carrière

Keteleeria davidiana (Bertrand) Beissn., Handb. Nadelholzk.: 424. Feb-Mar 1891. – Basionym: *Pseudotsuga davidiana* Bertrand in Ann. Sci. Nat., Bot., ser. 5, 20: 86-87. 1874. – Holotype: China, Sichuan, Longan-fou, "1870" [Dec 1869], *Abbé David 36* (P).

Keteleeria davidiana var. ***formosana*** (Hayata) Hayata in J. Coll. Sci. Imp. Univ. Tokyo 25(19): 221, f. 11. 23 Jul 1908. – Basionym and type: see under *Keteleeria formosana.*

Keteleeria evelyniana Mast. in Gard. Chron., ser. 3, 33: 194. 28 Mar 1903. – Holotype: China, Yunnan, near Yuan-chiang (Jianchuan), *A. Henry 11815* (NY).

Keteleeria formosana Hayata in Gard. Chron., ser. 3, 43: 194. 28 Mar 1908. – Holotype: Taiwan, Shinguki, Shinkocho, Nov 1902, *N. Konishi s.n.* (BM).

Keteleeria fortunei (A. Murray bis) Carrière in Rev. Hort. 37: 449. 1866. – Basionym: *Picea fortunei* A. Murray bis in Proc. Roy. Hort. Soc. London 2: 421, f. 85-97. 1 Jul 1862. – Lectotype (Farjon, 1989): China, Fujian, Fuzhou, (Foochow-foo), *R. Fortune 52* (BM).

Larix Mill.

Larix sect. ***Multiseriales*** Patschke in Bot. Jahrb. Syst. 48: 652, 770. 14 Jan 1913. – Lectotype (designated here): *Larix griffithii* Hook f. & Thomson [= *Larix griffithiana* (Lindl. & Gordon) Carrière].

Larix chinensis Beissn. in Mitt. Deutsch. Dendrol. Ges. 5: 68. 1896. – Type not designated.

Larix ×***czekanowskii*** Szafer in Kosmos (Lvov) 38: 1297. 1913 [& in Bot. Centralbl. 128: 29. 5 Jan 1915]. – Type not designated.

Larix decidua Mill., Gard. Dict., ed. 8, *Larix* No. 1. 16 Apr 1768. – Syn. subst.: *Pinus larix* L., Sp. Pl.: 1001. 1 Mai 1753. – Lectotype (designated here by A. Farjon & C. E. Jarvis): Soc. Gardeners [Miller], Cat. Pl.: t. 11. 1730.

Larix decidua subsp. ***polonica*** (Racib. ex Wóycicki) Domin in Acta Bot. Bohem. 10: 6. 1931. –

Basionym and type: see under *Larix polonica.*

Larix decidua var. ***carpatica*** Domin, Sborn. Vyzk. Ustav Zemed. R.Č.S. 65: 149. 1930. – Holotype: Czechoslovakia, Tatra Mts., 7 Sep 1931, *V. Krajina s.n.* (PR).

Larix decidua var. ***polonica*** (Racib. ex Wóycicki) Ostenf. & Syrach, Pflanzenareale 2: 63. 1930 [& in Biol. Meddel. Kongel. Danske Vidensk. Selsk. 9(2): 78. 30 Jun 1930]. – Basionym and type: see under *Larix polonica.*

Larix gmelinii (Rupr.) Rupr., Fl. Bor.-Ural.: 48. Mai 1854. – Basionym: *Abies gmelinii* Rupr. [Fl. Samojed. Cisural.] in Beitr. Pflanzenk. Russ. Reiches 2: 56. Jun 1845. – Type not designated.

Larix gmelinii var. ***japonica*** (Regel) Pilg. in Engler & Prantl, Nat. Pflanzenfam., ed. 2, 13: 327. 1926. – Basionym: *Larix dahurica* var. *japonica* Regel in Gartenflora 20: 105, t. 685, f. 6. Apr 1871. – Type not designated.

Larix gmelinii var. ***olgensis*** (A. Henry) Ostenf. & Syrach, Pflanzenareale 2: 62. 1930 [& in Biol. Meddel. Kongel. Danske Vidensk. Selsk. 9(2): 51. 30 Jun 1930]. – Basionym and type: see under *Larix olgensis.*

Larix gmelinii var. ***principisrupprechtii*** (Mayr) Pilg. in Engler & Prantl, Nat. Pflanzenfam., ed. 2, 13: 327. 1926. – Basionym and type: see under *Larix principisrupprechtii.*

Larix griffithiana (Lindl. & Gordon) Carrière, Traité Gén. Conif.: 278. Jan 1855. – Basionym: *Abies griffithiana* Lindl. & Gordon in J. Hort. Soc. London 5: 214. 1850. – Type not designated.

Larix griffithiana var. ***speciosa*** (W. C. Cheng & Y. W. Law) Silba in Phytologia 68: 37. Jan 1990 [& Farjon in Regnum Veg. 121: 211. Nov 1990]. – Basionym and type: see under *Larix speciosa.*

Larix himalaica W. C. Cheng & L. K. Fu in Acta Phytotax. Sin. 13(4): 84. Oct 1975. – Holotype: China, Xizang (Tibet), north side of Xomolungma (Mt. Everest), Chinese collector No. 80A (PE).

Larix kaempferi (Lamb.) Carrière, Fl. Serres Jard. Eur. 11: 97. 1856. – Basionym: *Pinus kaempferi* Lamb., Descr. Pinus 2: v. 1824. – Lectotype (Hara & Brummitt, 1980): Kaempfer, Delin. Pl. Japon.: t. 218 [ms.] in Bibl. Sloan., Min. 139, Cat. No. 2914 xxvi G (Brit. Mus., Bloomsbury).

Larix laricina (Du Roi) K. Koch, Dendrologie 2(2): 263. Nov 1873. – Basionym: *Pinus laricina* Du Roi, Diss. Inaug. Obs. Bot.: 49. 1771. – Type not designated.

Larix lyallii Parl., Conif. Nov.: 3. Jan 1863 [& in J. Bot. 1: 35. Mar 1863]. – Type not designated.

Larix mastersiana Rehder & E. H. Wilson in Sargent, Pl. Wilson. 2: 19. 24 Mar 1914. – Holotype: China, Sichuan, near Guanxian, 21 Jun 1908, *E. H. Wilson 906* (A; isot.: E, K).

Larix occidentalis Nutt., N. Amer. Sylva 3: 143, t. 120. 1849. – Type not designated.

Larix olgensis A. Henry in Gard. Chron., ser. 3, 57: 109, f. 31-32. 1915. – Holotype: U.S.S.R., Primor'ye, Ol'ga ("St. Olga"), 1860, *C. J. Maximowicz s.n.* (K).

Larix polonica Racib. ex Wóycicki, Obraz. Rośl. Król. Polsk. 2: 15-16, t. 1. 1912. – Type not designated.

Larix potaninii Batalin in Trudy Imp. S.-Peterburgsk. Bot. Sada 13: 385. 1894. – Type not designated.

Larix potaninii var. ***himalaica*** (W. C. Cheng & L. K. Fu) Farjon & Silba in Phytologia 68: 37. Jan 1990 [& Farjon in Regnum Veg. 121: 213. Nov 1990]. – Basionym and type: see under *Larix himalaica.*

Larix potaninii var. ***macrocarpa*** Y. W. Law in Acta Phytotax. Sin. 13(4): 84. Oct 1975. – Holotype: China, Yunnan, Chinese collector No. 9347 (PE).

Larix principis-rupprechtii Mayr, Fremdl. Wald- Parkbäume: 309, f. 94. Jan-Feb 1906. – Type not designated.

Larix sibirica Ledeb., Fl. Altaica 4: 204. Jul-Dec 1833. – Type not designated.

Larix speciosa W. C. Cheng & Y. W. Law in Acta Phytotax. Sin. 13(4): 84. Oct 1975. – Holotype: China, Yunnan, Wei-hsih, *Fung 20914* (PE).

Nothotsuga Hu ex C. N. Page

Nothotsuga longebracteata (W. C. Cheng) Hu ex C. N. Page in Notes Roy. Bot. Gard. Edinburgh 45: 390. 22 Feb 1988. – Basionym and type: see under *Tsuga longibracteata.*

Picea A. Dietrich

Picea sect. ***Casicta*** Mayr, Monogr. Abietin. Japan. Reich.: 44. Nov-Dec 1890. – Lectotype (designated here): *Picea ajanensis* Trautvetter & C. A. Mey. [= *Picea jezoënsis* (Siebold & Zucc.) Carrière].

Picea sect. ***Omorika*** Willk., Forstl. Fl. Deutschl. Oesterr., ed. 2: 66. Feb 1886. – Holotype: *Picea omorika* (Pančić) Purk.

Picea sect. ***Pungentes*** (Bobrov) P. A. Schmidt in Haussknechtia 4: 38. 13 Dec 1988. – Basionym: *Picea* ser. *Pungentes* Bobrov in Novosti Sist. Vysš. Rast. 7: 38. 4 Mai 1971. – Holotype: *Picea pungens* Engelm.

Picea subsect. ***Omorika*** (Willk.) E. Murray in Kalmia 14: 13. Jan 1984. – Basionym and type: see under *Picea* sect. *Omorika.*

Picea subsect. ***Pungentes*** (Bobrov) E. Murray in Kalmia 14: 14. Jan 1984. – Basionym and type: see under *Picea* sect. *Pungentes.*

Picea subsect. ***Sitchenses*** E. Murray in Kalmia 14: 14. Jan 1984. – Holotype: *Picea sitchensis* (Bong.) Carrière.

Picea ser. **Rubentes** Bobrov in Novosti Sist. Vysš. Rast. 7: 35. 4 Mai 1971. – Holotype: *Picea rubens* Sarg.

Picea abies (L.) H. Karst., Deut. Fl.: 324. Feb 1881. – Basionym: *Pinus abies* L., Sp. Pl.: 1002. 1 Mai 1753. – Lectotype (designated here by A. Farjon & C. E. Jarvis): illustration in Camerarius, Pl. Epit. Matthioli: 47. 1586.

Picea abies subsp. *obovata* (Ledeb.) Hultén in Svensk Bot. Tidskr. 43: 388. 8 Jul 1949. – Basionym and type: see under *Picea obovata*.

Picea abies var. *acuminata* (Beck) Dallim. & A. B. Jacks., Handb. Conif., ed. 3: 390. 1948. – Basionym: *Picea excelsa* var. *acuminata* Beck in Ann. K.K. Naturhist. Hofmus. 2: 39. Jan-Mar 1887. – Lectotype (designated here): Kienitz, Formen & Abarten Heim. Waldbäume: t. 3, f. 5a. 1879.

Picea abies var. *alpestris* (Brügger) P. A. Schmidt in Haussknechtia 4: 38. 13 Dec 1988. – Basionym: *Abies alpestris* Brügger in Jahresber. Naturf. Ges. Graubünden, ser. 2, 29: 167. 1886. – Type not designated.

Picea abies var. *obovata* (Ledeb.) Lindq. in Acta Horti Berg. 14: 307. 1948. – Basionym and type: see under *Picea obovata*.

Picea alcoquiana (Veitch ex Lindl.) Carrière, Traité Gén. Conif., ed. 2: 343. 15 Jan 1867. – Basionym: *Abies alcoquiana* Veitch ex Lindl. in Gard. Chron. 1861: 23. 12 Jan 1861. – Type not designated.

Picea alcoquiana var. *acicularis* (Shiras. & Koyama) Fitschen in Beissner, Handb. Nadelholzk., ed. 3: 258. Apr-Dec 1930. – Basionym: *Picea bicolor* var. *acicularis* Shiras. & Koyama in Bot. Mag. (Tokyo) 27: 130. 1913 [& in Mitt. Deutsch. Dendrol. Ges. 23: 254. 1914]. – Type not designated.

Picea alcoquiana var. *reflexa* (Shiras. & Koyama) Fitschen in Beissner, Handb. Nadelholzk., ed. 3: 258. Apr-Dec 1930. – Basionym: *Picea bicolor* var. *reflexa* Shiras. & Koyama in Bot. Mag. (Tokyo) 27: 130. 1913 [& in Mitt. Deutsch. Dendrol. Ges. 23: 255. 1914]. – Type not designated.

Picea asperata Mast. in J. Linn. Soc., Bot. 37: 419. 1 Nov 1906. – Holotype: China, Sichuan, *E. H. Wilson 3025* (BM).

Picea asperata var. *aurantiaca* (Mast.) Boom, Ned. Dendrol., ed. 10: 96. 1978. – Basionym and type: see under *Picea aurantiaca*.

Picea asperata var. *heterolepis* (Rehder & E. H. Wilson) Rehder, Man. Cult. Trees, ed. 2: 24. 1940. – Basionym: *Picea heterolepis* Rehder & E. H. Wilson in Sargent, Pl. Wilson. 2: 24. 24 Mar 1914. – Holotype: China, Sichuan, near Guanxian, *E. H. Wilson 4064* (A).

Picea asperata var. *ponderosa* Rehder & E. H. Wilson in Sargent, Pl. Wilson. 2: 23. 24 Mar 1914. – Holotype: China, Sichuan, W of Guanxian, *E. H. Wilson 4068* (A).

Picea asperata var. *retroflexa* (Mast.) W. C. Cheng in Chen, Taxon. Chin. Trees: 38. 1937. – Basionym and type: see under *Picea retroflexa*.

Picea aurantiaca Mast. in J. Linn. Soc., Bot. 37: 420. 1 Nov 1906. – Holotype: China, Sichuan, near Kangding (Tachien-lu), *E. H. Wilson 3029* (BM; isot.: E).

Picea balfouriana Rehder & E. H. Wilson in Sargent, Pl. Wilson. 2: 30. 24 Mar 1914. – Holotype: China, Sichuan, W of Kangding (Tachien-lu), *E. H. Wilson 4080* (A).

Picea brachytyla (Franch.) Pritz. in Bot. Jahrb. Syst. 29: 216. 4 Sep 1900. – Basionym: *Abies brachytyla* Franch. in J. Bot. (Morot) 13: 258. Aug 1899. – Type not designated.

Picea brachytyla var. *complanata* (Mast.) W. C. Cheng ex Rehder, Man. Cult. Trees, ed. 2: 30. 1940. – Basionym and type: see under *Picea complanata*.

Picea brachytyla var. *rhombisquamea* Stapf in Bot. Mag. 148: sub t. 8969. 13 Aug 1923. – Lectotype (designated here): China, Yunnan, *E. H. Wilson 3031* (K).

Picea breweriana S. Watson in Proc. Amer. Acad. Arts 20: 378. 21 Feb 1885. – Type not designated.

Picea chihuahuana Martínez in Anales Inst. Biol. Univ. Nac. México 13: 31. 1942. – Holotype: Mexico, Chihuahua, Arroyo de los Talayotes, 3 Feb 1942, *R. Dueñas s.n.* (MEXU).

Picea complanata Mast. in Gard. Chron., ser. 3, 39: 146. 10 Mar 1906. – Type not designated.

Picea crassifolia Kom. in Bot. Mater. Gerb. Glavn. Bot. Sada R.S.F.S.R. 4: 177. 31 Dec 1923. – Holotype: China, Qinghai, Qinghai Hu (Lake Kuku Nor), Qinghai Nanshan, near Dulanchit monastery, alt. 3300-3600 m, 14-26 Feb 1800, *N. M. Prshewalski s.n.* (LE).

Picea engelmannii Parry ex Engelm. in Trans. Acad. Sci. St. Louis 2: 212. 5 Mai 1863. – Lectotype (designated here): U.S.A., Wyoming, summit of Wind River Mountains, 7 Jun 1860, *F. V. Hayden s.n.* (MO).

Picea engelmannii subsp. *mexicana* (Martínez) P. A. Schmidt in Haussknechtia 4: 38. 13 Dec 1988. – Basionym and type: see under *Picea mexicana*.

Picea engelmannii var. *mexicana* (Martínez) R. J. Taylor & T. F. Patt. in Taxon 29: 438. 26 Aug 1980. – Basionym and type: see under *Picea mexicana*.

Picea farreri C. N. Page & Rushforth in Notes Roy. Bot. Gard. Edinburgh 38: 130. 26 Jun 1980. – Holotype: Burma, Feng-Shui-Ling Valley, *R. Farrer 1435* (E).

Picea ×*fennica* (Regel) Kom., Fl. S.S.S.R. 1: 145. Jan 1934. – Basionym: *Pinus abies* var. *fennica* Regel in Gartenflora 12: 95. Mar 1863. – Type not designated.

Picea glauca (Moench) Voss in Mitt. Deutsch. Dendrol. Ges. 16: 93. 1907. – Basionym: *Pinus glauca* Moench, Verz. Ausländ. Bäume: 73. 1785. – Type not extant.

Picea glauca subsp. *albertiana* (S. Br.) P. A. Schmidt in Haussknechtia 4: 38. 13 Dec 1988. – Basionym and type: see under *Picea glauca* var. *albertiana*.

Picea glauca var. *albertiana* (S. Br.) Sarg. in Bot. Gaz. 67: 208. 1919. – Basionym: *Picea albertiana* S. Br. in Torreya 7: 126. 1907. – Type not designated.

Picea glauca var. *densata* L. H. Bailey, Cult. Conif. N. Amer.: 108. 1933. – Type not designated.

Picea glehnii (F. Schmidt) Mast. in Gard. Chron., ser. 2, 13: 300. 6 Mar 1880. – Basionym: *Abies glehnii* F. Schmidt [Reis. Amur-Lande, Bot.] in Mém. Acad. Imp. Sci. Saint Pétersbourg, ser. 7, 12(2): 176. Sep-Oct 1868. – Holotype: U.S.S.R., S. Sakhalin, Aug 1861, *P. von Glehn s.n.* (LE).

Picea hirtella Rehder & E. H. Wilson in Sargent, Pl. Wilson. 2: 32. 24 Mar 1914. – Holotype: China, Sichuan, Panlan Shan, *E. H. Wilson 2084* (A).

Picea hondoënsis Mayr, Monogr. Abietin. Japan. Reich.: 51, t. 4, f. 9. Nov-Dec 1890. – Type not designated.

Picea jezoënsis (Siebold & Zucc.) Carrière, Traité Gén. Conif.: 255. Jan 1855. – Basionym: *Abies jezoënsis* Siebold & Zucc., Fl. Jap.

2: 19, t. 110. 1842. – Lectotype (designated here): Japan, "Abies No. 2, Coniferae ex insula Jezo", *P. F. von Siebold* (L).

Picea jezoënsis subsp. *hondoënsis* (Mayr) P. A. Schmidt in Haussknechtia 4: 38. 13 Dec 1988. – Basionym and type: see under *Picea hondoënsis*.

Picea jezoënsis var. *hondoënsis* (Mayr) Rehder in Mitt. Deutsch. Dendrol. Ges. 24: 314. 1915. – Basionym and type: see under *Picea hondoënsis*.

Picea jezoënsis var. *komarovii* (V. N. Vassil.) W. C. Cheng & L. K. Fu in Fl. Reipubl. Pop. Sin. 7: 161. Dec 1978. Basion.: *Picea komarovii* V. N. Vassil. in Bot. Žurn. (Moscow & Leningrad) 35: 504, f. 5, 7. 23 Oct 1950. – Type not designated.

Picea koraiensis Nakai in Bot. Mag. (Tokyo) 33: 195. 1919 [& in J. Jap. Bot. 2: 9. 1919]. – Type not designated.

Picea koraiensis var. *intercedens* (Nakai) Y. L. Chou, Lign. Fl. Heilongjiang: 49. 1986. – Basionym: *Picea intercedens* Nakai in J. Jap. Bot. 17(1): 4. Jan 1941. – Type not designated.

Picea koraiensis var. *pungsanensis* (Uyeki ex Nakai) Schmidt-Vogt ex Farjon in Regnum Veg. 121: 231. Nov 1990. – Basionym: *Picea pungsanensis* Uyeki ex Nakai in J. Jap. Bot. 17(1): 3. Jan 1941. – Type not designated.

Picea koyamae Shiras. in Bot. Mag. (Tokyo) 27: 127. 1913 [& in Mitt.

Deutsch. Dendrol. Ges. 23: 254. 1914]. – Type not designated.

Picea likiangensis (Franch.) Pritz. in Bot. Jahrb. Syst. 29: 217. 4 Sep 1900. – Basionym: *Abies likiangensis* Franch. in J. Bot. (Morot) 13: 257. Aug 1899. – Holotype: China, Yunnan, *P. J. M. Delavay 1031* (P).

Picea likiangensis var. *hirtella* (Rehder & E. H. Wilson) W. C. Cheng in Chen, Taxon. Chin. Trees: 40. 1937. – Basionym and type: see under *Picea hirtella*.

Picea likiangensis var. *linzhiensis* W. C. Cheng & L. K. Fu in Acta Phytotax. Sin. 13(4): 83. Oct 1975. – Holotype: China, Xizang (Tibet), Linzhi, 3100 m, Chinese collector No. 676 (PE).

Picea likiangensis var. *montigena* (Mast.) W. C. Cheng in Chen, Taxon. Chin. Trees: 40. 1937. – Basionym and type: see under *Picea montigena*.

Picea likiangensis var. *purpurea* (Mast.) Dallim. & A. B. Jacks., Handb. Conif.: 334. Sep-Dec 1923. – Basionym and type: see under *Picea purpurea*.

Picea likiangensis var. *rubescens* Rehder & E. H. Wilson in Sargent, Pl. Wilson. 2: 31. 24 Mar 1914. – Holotype: China, Sichuan, near Kangding (Tachienlu), *E. H. Wilson 2057* (A).

Picea ×*lutzii* Little in J. Forest. (Washington) 51: 745. 1953. – Type not designated.

Picea mariana (Mill.) Britton & al., Prelim. Cat.: 71. 25 Apr 1888. –

Basionym: *Abies mariana* Mill., Gard. Dict., ed. 8, *Abies* No. 5. 16 Apr 1768. – Type not designated.

Picea martinezii T. F. Patt. in Sida 13: 131. 30 Dec 1988. – Holotype: Mexico, Nuevo León, S.E. of La Trinidad, *T. F. Patterson 5629* (TEX; isot.: MEXU, MO, US).

Picea maximowiczii Regel ex Mast. in Gard. Chron., ser. 2, 11: 363. 22 Mar 1879. – Type not designated.

Picea maximowiczii var. *senanensis* Hayashi, Ill. Useful Trees (Forest Trees): sub f. 43. 1969. – Type not designated.

Picea mexicana Martínez in Anales Inst. Biol. Univ. Nac. México 32: 137. 1961. – Holotype: Mexico, Nuevo León, El Carmen, 26 Jun 1961, *J. Vazquez 500a* (MEXU).

Picea meyeri Rehder & E. H. Wilson in Sargent, Pl. Wilson. 2: 28. 24 Mar 1914. – Holotype: China, Shanxi, Wutai Shan, *F. N. Meyer 22672* (A).

Picea montigena Mast. in Gard. Chron., ser. 3, 39: 146. 10 Mar 1906. – Holotype: China, Sichuan, near Kangding (Tachienlu), *E. H. Wilson 3027* (BM).

Picea morrisonicola Hayata in J. Coll. Sci. Imp. Univ. Tokyo 25(19): 220. 23 Jul 1908. – Lectotype (designated here): Taiwan, Mt. Morrison, Nov 1906, *T. Kawakami & U. Mori 2108* (TI).

Picea neoveitchii Mast. in Gard. Chron., ser. 3, 33: 116, f. 50, 51. 21 Feb 1903. – Holotype: China, Hubei, *E. H. Wilson 2601* (BM).

Picea ×*notha* Rehder in J. Arnold Arbor. 20: 85. 26 Jan 1939. – Holotype: from cultivated tree at Arnold Arboretum, *A. Rehder & E. J. Palmer (13406)* (A).

Picea obovata Ledeb., Fl. Altaica 4: 201. Jul-Dec 1833. – Type not designated.

Picea obovata var. *fennica* (Regel) A. Henry in Elwes & Henry, Trees Great Britain: 1360. Jan 1913. – Basionym and type: see under *Picea* ×*fennica*.

Picea omorika (Pančić) Purk. in Österr. Monatsschr. Forstwesen 27: 446. 1877. – Basionym: *Pinus omorika* Pančić, Neue Conif. Alp.: 4. 1876 [& in Gard. Chron., ser. 2, 7: 620. 19 Mai 1877]. – Type not designated.

Picea orientalis (L.) Link in Linnaea 20: 294. Jun 1847. – Basionym: *Pinus orientalis* L., Sp. Pl., ed. 2: 1421. Jul-Aug 1763. – Type not designated.

Picea pungens Engelm. in Gard. Chron., ser. 2, 11: 334. 15 Mar 1879. – Lectotype (designated here): U.S.A., Colorado, Clear Creek, 4 Sep 1874, *G. Engelmann s.n.* (MO).

Picea purpurea Mast. in J. Linn. Soc., Bot. 37: 418. 1 Nov 1906. – Holotype: China, Sichuan, *E. H. Wilson 3026* (BM).

Picea retroflexa Mast. in J. Linn. Soc., Bot. 37: 420. 1 Nov 1906. – Holotype: China, Sichuan, near Kangding (Tachien-lu), *E. H. Wilson 3030* (A).

Picea rubens Sarg., Silva 12: 33. 10 Jan 1899. – Type not designated.

Picea schrenkiana Fisch. & C. A. Mey. in Bull. Sci. Acad. Imp. Sci. Saint-Pétersbourg 10: 253. 1842. – Type not designated.

Picea schrenkiana subsp. *tianschanica* (Rupr.) Bykov in Izv. Akad. Nauk Kazahsk. S.S.R., Ser. Bot. 5: 22. 1950. – Basionym and type: see under *Picea tianschanica*.

Picea schrenkiana var. *tianschanica* (Rupr.) W. C. Cheng & S. H. Fu in Fl. Reipubl. Pop. Sin. 7: 146. Dec 1978. – Basionym and type: see under *Picea tianschanica*.

Picea shirasawae Hayashi, Ill. Useful Trees (Forest Trees): sub f. 43. 1969. – Syn. subst. and type: see under *Picea alcoquiana* var. *acicularis*.

Picea sitchensis (Bong.) Carrière, Traité Gén. Conif.: 260. Jan 1855. – Basionym: *Pinus sitchensis* Bong. in Mém. Acad. Imp. Sci. St.-Pétersbourg, ser. 6, Sci. Math. 2: 164. 1832. – Type not designated.

Picea smithiana (Wall.) Boiss., Fl. Orient. 5: 700. Apr 1884. – Basionym: *Pinus smithiana* Wall., Pl. Asiat. Rar. 3: 24, f. 246. 20 Mar 1832. – Type not designated.

Picea spinulosa (Griff.) A. Henry in Gard. Chron., ser. 3, 39: 219, f. 84. 7 Apr 1906. – Basionym: *Abies spinulosa* Griff., Itin. Pl. Khasyah Mts.: 259. 1848. – Type not designated.

Picea tianschanica Rupr. in Mém. Acad. Imp. Sci. Saint Pétersbourg, ser. 7, 14(4): 72. 1869. – Type not designated.

Picea torano (Siebold ex K. Koch) Koehne, Deut. Dendrol.: 22. Mai 1893. – Basionym: *Abies torano* Siebold ex K. Koch, Dendrologie 2(2): 233. Nov 1873. – Lectotype (designated here): Japan, "Owari pr., Abies No. 1 torano ki", *P. F. von Siebold* sub *Abies polita* (L).

Picea wilsonii Mast. in Gard. Chron., ser. 3, 33: 133. 28 Feb 1903. – Holotype: China, Hubei, *E. H. Wilson 1897* (BM).

Pinus L.

Pinus subg. *Ducampopinus* (A. Cheval.) Little & Critchf. [Geogr. Distr. Pines] in Misc. Publ. U.S. Dept. Agric. 991: 5. Feb 1966. – Basionym: *Ducampopinus* A. Cheval. in Rev. Int. Bot. Appl. Agric. Trop. 24: 30. 1944. – Holotype: *Ducampopinus krempfii* (Lecomte) A. Cheval. (*Pinus krempfii* Lecomte).

Pinus subg. *Strobus* (Sweet ex Spach) Lemmon, Cone-Bear. Trees Pacif. Slope, ed. 3: 20. Jul 1895. – Holotype: *Pinus strobus* L.

Pinus sect. *Leiophyllae* Burgh in Rev. Palaeobot. Palynol. 15(2-3): 92. 1973. – Holotype: *Pinus leiophylla* Schiede ex Schltdl. & Cham.

Pinus sect. *Lumholtziae* Burgh in Rev. Palaeobot. Palynol. 15(2-3): 92. 1973. – Holotype: *Pinus lumholtzii* B. L. Rob. & Fernald.

Pinus sect. *Parryanae* Mayr, Wald. Nordamer. 241, 427. Sep-Nov 1889. – Holotype: *Pinus parryana* Engelm. [= *Pinus quadrifolia* Parl. ex Sudw.].

Pinus sect. *Pinea* Endl., Syn. Conif.: 182. Mai-Jun 1847. – Holotype: *Pinus pinea* L.

Pinus sect. *Strobus* Sweet ex Spach, Hist. Nat. Vég. 11: 394. 25 Dec 1841. – Holotype: *Pinus strobus* L.

Pinus subsect. *Aristatae* Burgh in Rev. Palaeobot. Palynol. 15(2-3): 90. 1973. – Holotype: *Pinus aristata* Engelm.

Pinus subsect. *Attenuatae* Burgh in Rev. Palaeobot. Palynol. 15(2-3): 93. 1973. – Holotype: *Pinus attenuata* Lemmon.

Pinus subsect. *Australes* Loudon, Arbor. Frutic. Brit.: 2255. 1 Jul. 1838. – Holotype: *Pinus australis* Michx. [= *Pinus palustris* Mill.].

Pinus subsect. *Balfourianae* Engelm. [Rev. Pinus] in Trans. Acad. Sci. St. Louis 4: 176. Feb 1880. – Holotype: *Pinus balfouriana* Balf. ex A. Murray bis.

Pinus subsect. *Canarienses* Loudon, Arbor. Frutic. Brit.: 2261. 1 Jul 1838. – Holotype: *Pinus canariensis* C. Sm.

Pinus subsect. *Cembroides* Engelm. [Rev. Pinus] in Trans. Acad. Sci. St. Louis 4: 178. Feb 1880. – Holotype: *Pinus cembroides* Zucc.

Pinus subsect. *Contortae* Little & Critchf. [Geogr. Distr. Pines] in Misc. Publ. U.S. Dept. Agric. 991:

19. Feb 1966. – Holotype: *Pinus contorta* Douglas ex Loudon.

Pinus subsect. *Gerardianae* Loudon, Arbor. Frutic. Brit.: 2254. 1 Jul 1838. – Holotype: *Pinus gerardiana* Wall. ex D. Don.

Pinus subsect. *Halepenses* Burgh in Rev. Palaeobot. Palynol. 15(2-3): 92. 1973. – Holotype: *Pinus halepensis* Mill.

Pinus subsect. *Krempfianae* Little & Critchf. [Geogr. Distr. Pines] in Misc. Publ. U.S. Dept. Agric. 991: 5. Feb 1966. – Holotype: *Pinus krempfii* Lecomte.

Pinus subsect. *Nelsoniae* Burgh in Rev. Palaeobot. Palynol. 15(2-3): 92. 1973. – Holotype: *Pinus nelsonii* Shaw.

Pinus subsect. *Oocarpae* Little & Critchf. [Geogr. Distr. Pines] in Misc. Publ. U.S. Dept. Agric. 991: 19. Feb 1966. – Holotype: *Pinus oocarpa* Schiede ex Schltdl.

Pinus subsect. *Pinea* (Endl.) Little & Critchf. [Subdiv. Gen. Pinus] in Misc. Publ. U.S. Dept. Agric. 1144: 12. Dec 1969. – Holotype: *Pinus pinea* L.

Pinus subsect. *Ponderosae* Loudon, Arbor. Frutic. Brit.: 2243. 1 Jul 1838. – Holotype: *Pinus ponderosa* Douglas ex Lawson.

Pinus subsect. *Pseudostrobus* Burgh in Rev. Palaeobot. Palynol. 15(2-3): 93. 1973. – Holotype: *Pinus pseudostrobus* Lindl.

Pinus subsect. *Rzedowskianae* Carvajal in Phytologia 59: 134. 11 Jan 1986. – Holotype: *Pinus rzedowskii* Madrigal & M. Caball.

Pinus subsect. *Sabinianae* Loudon, Arbor. Frutic. Brit.: 2246. 1 Jul 1838. – Holotype: *Pinus sabiniana* Douglas ex D. Don.

Pinus subsect. *Strobi* Loudon, Arbor. Frutic. Brit.: 2280. 1 Jul 1838. – Holotype: *Pinus strobus* L.

Pinus subsect. *Torreyanae* Burgh in Rev. Palaeobot. Palynol. 15(2-3): 94. 1973. – Holotype: *Pinus torreyana* Parry ex Carrière.

Pinus albicaulis Engelm. in Trans. Acad. Sci. St. Louis 2: 209. 5 Mai 1863. – Holotype: U.S.A., Colorado, *Newberry s.n.* (US No. 61140).

Pinus amamiana Koidz. in Bot. Mag. (Tokyo) 38: 113. 1924. – Type not designated.

Pinus apulcensis Lindl., Edwards's Bot. Reg. 25: 63. Aug 1839 [& in Allg. Gartenzeitung 7: 325. 1839]. – Type not designated.

Pinus aristata Engelm. in Amer. J. Sci. Arts, ser. 2, 34: 331. 1862. – Lectotype (designated here): U.S.A., Colorado, Jul 1862 *C. C. Parry s.n.* (MO).

Pinus aristata var. *longaeva* (D. K. Bailey) Little in Phytologia 42: 221. 31 Mar 1979. – Basionym and type: see under *Pinus longaeva*.

Pinus arizonica Engelm. in Rothrock, Rep. U.S. Geogr. Surv., Wheeler, 6, Bot.: 260. Jun-Aug 1879. – Holotype: U.S.A., Arizona, *J. T. Rothrock 652* (MO).

Pinus arizonica var. *stormiae* Martínez, Pinos Mexic., ed. 2: 295.

1948. – Holotype: Mexico, Coahuila, *M. Martínez 3455* (MEXU).

Pinus armandii Franch. in Nouv. Arch. Mus. Hist. Nat. ser. 2, 7: 93, t. 7. Dec 1884 [& Pl. David. 1: 283. 1884]. – Type not designated.

Pinus armandii var. ***dabeshanensis*** (W. C. Cheng & Y. W. Law) Silba in Phytologia 68: 47. Jan 1990. – Basionym and type: see under *Pinus dabeshanensis.*

Pinus armandii var. ***mastersiana*** (Hayata) Hayata in J. Coll. Sci. Imp. Univ. Tokyo 25(19): 217, f. 8. 23 Jul 1908. – Basionym: *Pinus mastersiana* Hayata in Gard. Chron., ser. 3, 43: 194. 28 Mar 1908. – Holotype: Taiwan, Mt. Morrison, 3 Oct 1905, *G. Nakahara s.n.* (TI).

Pinus attenuata Lemmon in Mining Sci. Press 64: 45. 16 Jan 1892. – Lectotype (designated here): U.S.A., California, *J. G. Lemmon s.n.* (UC No. 338321).

Pinus ayacahuite Ehrenb. ex Schltdl. in Linnaea 12: 492. Apr-Sep 1838. – Type not designated.

Pinus ayacahuite var. ***brachyptera*** Shaw [Pines Mexico] in Publ. Arnold Arbor. 1: 10. Mar 1909. – Lectotype (designated here): Mexico, Durango, El Salto, *E. W. Nelson 4555* (A).

Pinus ayacahuite var. ***veitchii*** (Roezl) Shaw [Pines Mexico] in Publ. Arnold Arbor. 1: 10. Mar 1909. – Basionym: *Pinus veitchii* Roezl, Cat. Grain. Conif. Mexic.: 32. Jun 1857 [& Schltdl. in Linnaea 29: 353. Sep 1858]. – Type not designated.

Pinus balfouriana Balf. ex A. Murray bis, Bot. Exped. Oregon 8: 1. 1853. – Holotype: U.S.A., California, Mts. between Shasta and Scots Valley, *J. Jeffrey "618"* (E).

Pinus balfouriana subsp. ***austrina*** R. Mastrogiuseppe & J. Mastrogiuseppe in Syst. Bot. 5: 102. 31 Oct 1980. – Holotype: U.S.A., California, 8 Dec, *R. J. Mastrogiuseppe* (WS).

Pinus balfouriana var. ***austrina*** (R. Mastrogiuseppe & J. Mastrogiuseppe) Silba in Phytologia Mem. 7: 48. 1984. – Basionym and type: see under *Pinus balfouriana* subsp. *austrina.*

Pinus banksiana Lamb., Descr. Pinus 1: 7, t. 3. Sep 1803. – Lectotype (designated here): Lambert, Descr. Pinus 1: t. 3. Sep 1803.

Pinus bhutanica Grierson & al. in Notes Roy. Bot. Gard. Edinburgh 38: 299. 17 Dec 1980. – Holotype: Bhutan, 4 km N. of Tschilingor, *A. J. C. Grierson & D. G. Long 2282* (E; isot.: A, BM, K).

Pinus brutia Ten., Fl. Napol. 1: lix. 1811-1815. – Type not designated.

Pinus brutia subsp. ***eldarica*** (Medw.) Nahal in Ann. Ecole Natl. Eaux 19: 521. 1962. – Basionym and type: see under *Pinus eldarica.*

Pinus brutia subsp. ***pityusa*** (Steven) Nahal in Ann. Ecole Natl.

Eaux 19: 521. 1962. – Basionym and type: see under *Pinus pityusa.*

Pinus brutia var. ***eldarica*** (Medw.) Silba in Phytologia 58: 367. 20 Nov 1985. – Basionym and type: see under *Pinus eldarica.*

Pinus brutia var. ***pityusa*** (Steven) Silba in Phytologia 58: 367. 20 Nov 1985. – Basionym and type: see under *Pinus pityusa.*

Pinus brutia var. ***stankewiczii*** Sukaczev in Bot. Žurn. (St. Petersburg) 1: 37. 1906. – Type not designated.

Pinus bungeana Zucc. ex Endl., Syn. Conif.: 166. Mai-Jun 1847. – Type not designated.

Pinus californiarum D. K. Bailey in Notes Roy. Bot. Gard. Edinburgh 44: 278. 11 Jun 1987. – Holotype: U.S.A., California, San Diego County, *D. K. Bailey 81-08* (COLO; isot.: E, K, MEXU, MO, NY, UC, US).

Pinus californiarum subsp. ***fallax*** (Little) D. K. Bailey in Notes Roy. Bot. Gard. Edinburgh 44: 279. 11 Jun 1987. – Basionym and type: see under *Pinus edulis* var. *fallax.*

Pinus canariensis C. Sm. in Buch, Phys. Beschr. Canar. Ins.: 159. Jun-Dec 1828. – Type not extant.

Pinus caribaea Morelet in Rev. Hort. Côte d'Or 1: 107. 1851. – Type not designated.

Pinus caribaea var. ***bahamensis*** (Griseb.) W. H. Barrett & Golfari in Caribbean Forest. 23: 69. 1962. – Basionym: *Pinus bahamensis* Griseb., Fl. Brit. W.I.: 503. Mai 1862. – Neotype (Barrett & Golfari, 1962): Bahamas, *L. Golfari 77571* (BAB).

Pinus caribaea var. ***hondurensis*** (Sénéclauze) W. H. Barrett & Golfari in Caribbean Forest. 23: 65. 1962. – Basionym: *Pinus hondurensis* Sénéclauze, Conif.: 126. 1868. – Neotype (Barrett & Golfari, 1962): Honduras, *W. H. G. Barrett 77582* (BAB).

Pinus cembra L., Sp. Pl.: 1000. 1 Mai 1753. – Lectotype (designated here by A. Farjon & C. E. Jarvis): Breyne [De Balsamo Carpath.] in Acad. Caes.-Leop. Carol. Nat. Cur. Ephem. 7: t. 1. 1719.

Pinus cembra subsp. ***sibirica*** (Du Tour) Krylov, Fl. Zap. Sibir. 1: 77. 1927. – Basionym and type: see under *Pinus sibirica.*

Pinus cembra var. ***sibirica*** (Du Tour) G. Don in Loudon, Hort. Brit. 1: 387. 1830. – Basionym and type: see under *Pinus sibirica.*

Pinus cembroides Zucc. in Abh. Math.-Phys. Cl. Königl. Bayer. Akad. Wiss. 1: 392. 1832. – Holotype: Mexico, Hidalgo, near Zimapán, Sep 1827, *W. F. Karwinski* (M).

Pinus cembroides subsp. ***lagunae*** (Rob.-Pass.) D. K. Bailey in Phytologia 54: 98. 22 Oct 1983. – Basionym and type: see under *Pinus cembroides* var. *lagunae.*

Pinus cembroides subsp. ***orizabensis*** D. K. Bailey in Phytologia 54: 89. 22 Oct 1983. – Holotype: Mexico, Puebla, Soltepec, *D. K. Bailey 83-01* (MEXU).

Pinus cembroides var. *bicolor* Little in Phytologia 17: 331. 18 Oct 1968. – Holotype: U.S.A., New Mexico, *E. L. Little 23011* (US).

Pinus cembroides var. *lagunae* Rob.-Pass. in Bull. Mus. Natl. Hist. Nat., B, Adansonia 1: 64. 30 Jul 1981. – Holotype: Mexico, Baja California, Sierra de Laguna, *M.-F. Robert 10021* (P; isot.: MPU, TLJ, ENCB).

Pinus cembroides var. *quadrifolia* (Parl. ex Sudw.) de Laubenf. & Silba in Phytologia Mem. 7: 49. 1984. – Basionym and type: see under *Pinus quadrifolia.*

Pinus cembroides var. *remota* Little in Wrightia 3: 183. 31 Mai 1966. – Holotype: U.S.A., Texas, *E. L. Little & D. S. Correll 18991* (US).

Pinus chiapensis (Martínez) Andresen in Phytologia 10: 417. 30 Sep 1964. – Basionym and type: see under *Pinus strobus* var. *chiapensis.*

Pinus chihuahuana Engelm. in Wislizenus, Mem. Tour N. Mexico: 103. Apr 1848. – Type not designated.

Pinus clausa (Chapm. ex Engelm.) Sarg., Rep. For. N. America: 199. Sep-Dec 1884. – Basionym: *Pinus inops* var. *clausa* Chapm. ex Engelm. in Bot. Gaz. 2: 125. 1877. – Type not designated.

Pinus clausa var. *immuginata* D. B. Ward in Castanea 28: 4. 30 Mar 1963. – Holotype: U.S.A., Flo-rida, *D. B. Ward & Smith 2570* (FLAS).

Pinus contorta Douglas ex Loudon, Arbor. Frutic. Brit.: 2292. 1 Jul 1838. – Type not designated.

Pinus contorta subsp. *bolanderi* (Parl.) Koehne, Deut. Dendrol.: 37. Mai 1893. – Basionym: *Pinus bolanderi* Parl. in Candolle, Prodr. 16(2): 379. med Jul 1868. – Type not designated.

Pinus contorta subsp. *latifolia* (Engelm.) Critchf. in Publ. Maria Moors Cabot Found. Bot. Res. 3: 107. 1957. – Basionym and type: see under *Pinus contorta* var. *latifolia.*

Pinus contorta subsp. *murrayana* (A. Murray bis) Critchf. in Publ. Maria Moors Cabot Found. Bot. Res. 3: 106. 1957. – Basionym and type: see under *Pinus contorta* var. *murrayana.*

Pinus contorta var. *bolanderi* Lemmon in Erythea 2: 176. 1894. – Basionym and type: see under *Pinus contorta* subsp. *bolanderi.*

Pinus contorta var. *latifolia* Engelm. in Watson, Botany (Fortieth Parallel): 331. Sep-Dec 1871. – Type not designated.

Pinus contorta var. *murrayana* (A. Murray bis) Engelm. in Watson, Bot. California 2: 126. Jul-Dec 1880. – Basionym: *Pinus murrayana* A. Murray bis, Bot. Exped. Oregon 8: 2. 1853. – Type not designated.

Pinus cooperi C. E. Blanco in Anales Inst. Biol. Univ. Nac. México

20: 185. 1950. – Type not designated.

Pinus cooperi var. ***ornelasii*** C. E. Blanco in Anales Inst. Biol. Univ. Nac. México 20: 185. 1950. – Holotype: Mexico, Durango, El Salto, *M. Martínez 3441* (MEXU).

Pinus coulteri D. Don in Trans. Linn. Soc. London 17: 440. 21 Jun - 3 Jul 1836. – Type not designated.

Pinus cubensis Griseb. [Pl. Wright.] in Mem. Amer. Acad. Arts, ser. 2, 8: 530. Nov 1862. – Type not designated.

Pinus culminicola Andresen & Beaman in J. Arnold Arbor. 42: 437. 13 Oct 1961. – Holotype: Mexico, Nuevo León, Cerro Potosi, *J. H. Beaman 2675* (MSC; isot.: A, US).

Pinus dabeshanensis W. C. Cheng & Y. W. Law in Acta Phytotax. Sin. 13(4): 85. Oct 1975. – Holotype: China, Anhui, Dabie Shan, *C. C. Chang 1* (PE).

Pinus dalatensis Ferré in Bull. Soc. Hist. Nat. Toulouse 95: 178. 1960 [& in Trav. Lab. Forest. Toulouse 1(6, 4): 8. 1960]. – Holotype: Vietnam, Dalat, Trai Mat, *H. Gaussen s.n.* (TLF).

Pinus densa (Little & K. W. Dorman) Laubenfels & Silba in Phytologia Mem. 7: 50. 1984. – Basionym and type: see under *Pinus elliottii* var. *densa*.

Pinus densata Mast. in J. Linn. Soc., Bot. 37: 416. 1 Nov 1906. – Holotype: China, N.W. Yunnan, *E. H. Wilson 3015* (BM).

Pinus densiflora Siebold & Zucc., Fl. Jap. 2: 22, t. 112. 1842. – Lectotype (designated here): "in Japonia", *P. F. von Siebold* comm. 1842 ex herb. Zuccarini No. 438 (M).

Pinus densiflora var. ***funebris*** (Kom.) Liou & Wang in Liou, Ill. Fl. Lign. Pl. N.E. China: 98, 546. 1958. – Basionym: *Pinus funebris* Kom. [Fl. Manshur. 1] in Trudy Imp. S.-Peterburgsk. Bot. Sada 20: 177. Dec 1901. – Type not designated.

Pinus devoniana Lindl., Edwards's Bot. Reg. 25: 62. Aug 1839 [& in Allg. Gartenzeitung 7: 324. 1839]. – Type not designated.

Pinus discolor D. K. Bailey & Hawksw. in Phytologia 44: 130. 13 Sep 1979. – Holotype: U.S.A., New Mexico, *E. L. Little 23011* (US; isot.: A, NY).

Pinus donnell-smithii Mast. in Bot. Gaz. 16: 199. 1891. – Holotype: Guatemala, Volcán de Agua, alt. 4100 m, Apr 1890, *J. D. Smith 2182* (MO or US).

Pinus douglasiana Martínez in Madroño 7: 4. 28 Jan 1943. – Holotype: Mexico, *M. Martínez 3429* (MEXU).

Pinus durangensis Martínez in Anales Inst. Biol. Univ. Nac. México 13: 23, f. 1-4. 1942. – Type not designated.

Pinus echinata Mill., Gard. Dict., ed. 8: *Pinus* No. 12. 16 Apr 1768. – Type not designated.

Pinus edulis Engelm. in Wislizenus, Mem. Tour N. Mexico: 88. Apr 1848. – Type not designated.

Pinus edulis var. ***fallax*** Little in Phytologia 17: 331. 18 Oct 1968. – Holotype: U.S.A., Arizona, Tonto National Forest, *E. L. Little 18581* (US; isot.: A, NY).

Pinus eldarica Medw. in Trudy Tiflissk. Bot. Sada 6(2): 21. 1903. – Lectotype (designated here): illustration in Trudy Tiflissk. Bot. Sada 6(2): 21. 1903.

Pinus elliottii Engelm. [Rev. Pinus] in Trans. Acad. Sci. St. Louis 4: 186, t. 1-3. Feb 1880. – Lectotype (designated here): U.S.A., South Carolina, 20 Mar 1873, *J. H. Mellichamp s.n.* (MO No. 3941460).

Pinus elliottii var. ***densa*** Little & K. W. Dorman in J. Forest. (Washington) 50: 921, f. 1, 2. 1952. – Type not designated.

Pinus engelmannii Carrière in Rev. Hort., ser. 4, 3: 227. 1854. – Type not designated.

Pinus estevezii (Martínez) J. P. Perry in J. Arnold Arbor. 63: 187. 13 Apr 1982. – Basionym and type: see under *Pinus pseudostrobus* var. *estevezii*.

Pinus fenzeliana Hand.-Mazz. in Österr. Bot. Z. 80: 337. 28 Dec 1931. – Holotype: China, Hainan, 1929, *Fenzel 55* (WU).

Pinus flexilis E. James, Account Exped. Pittsburgh 2: 27, 35. 1823. – Neotype (designated here): U.S.A., Colorado, *J. W. Andresen & Barger A2125* (SIU).

Pinus flexilis var. ***reflexa*** Engelm. in Rothrock, Rep. U.S. Geogr. Surv., Wheeler, 6, Bot.: 258. Jun-Aug 1879. – Type not designated.

Pinus funebris Kom. [Fl. Manshur. 1] in Trudy Imp. S.-Peterburgsk. Bot. Sada 20: 177. Dec 1901. – Type not designated.

Pinus gerardiana Wall. ex D. Don in Lambert, Descr. Pinus, ed. 8°, 2: p. s.n. inter 144 et 145, t. 79. 1832. – Type not designated.

Pinus glabra Walter, Fl. Carol.: 237. Apr-Jun 1788. – Type not designated.

Pinus gordoniana Hartw. ex Gordon in J. Hort. Soc. London 2: 79. 1847. – Type not designated.

Pinus greggii Engelm. ex Parl. in Candolle, Prodr. 16(2): 396. med Jul 1868. – Type not designated.

Pinus halepensis Mill., Gard. Dict., ed. 8: *Pinus* No. 8. 16 Apr 1768. – Type not designated.

Pinus hartwegii Lindl., Edwards's Bot. Reg. 25: 62. Aug 1839 [& in Allg. Gartenzeitung 7: 324. 1839]. – Type not designated.

Pinus heldreichii H. Christ in Verh. Naturf. Ges. Basel 3: 549. 1863. – Type not designated.

Pinus heldreichii var. ***leucodermis*** (Antoine) Markgr. ex Fitschen in Beissner, Handb. Nadelholzk., ed. 3: 404. Apr-Dec 1930. – Basionym and type: see under *Pinus leucodermis*.

Pinus henryi Mast. in J. Linn. Soc., Bot. 26: 550. 21 Oct 1902. –

Holotype: China, Hubei, *A. Henry 6909* (K).

Pinus herrerae Martínez in Anales Inst. Biol. Univ. Nac. México 11: 76. 1940. – Type not designated.

Pinus hwangshanensis W. Y. Hsia in Chin. J. Bot. 1(1): 17, t. 6, f. 1. 1936. – Type not designated.

Pinus insularis Endl., Syn. Conif.: 157. Mai-Jun 1847. – Type not designated.

Pinus jaliscana Pérez de la Rosa in Phytologia 54: 290. 1 Dec 1983. – Holotype: Mexico, Jalisco, *J. A. Pérez de la Rosa 370* (IBUG).

Pinus jeffreyi Balf. ex A. Murray bis, Bot. Exped. Oregon 8: 2. 1853. – Type not designated.

Pinus johannis Rob.-Pass. in Adansonia, ser. 2, 18: 366. 28 Dec 1978. – Holotype: Mexico, Zacatecas, *M.-F. Robert 5936* (P).

Pinus juarezensis Lanner in Southw. Naturalist 19(1): 77. 1974. – Holotype: Mexico, Baja California, Sierra Juárez, *R. M. Lanner 2016* (UTC).

Pinus kesiya Royle ex Gordon in Gard. Mag. & Reg. Rural Domest. Improv. 16: 8. 1840. – Type not designated.

Pinus kesiya var. ***langbianensis*** (A. Cheval.) W. C. Cheng & L. K. Fu in Fl. Reipubl. Pop. Sin. 7: 259. Dec 1978. – Basionym: *Pinus langbianensis* A. Cheval. in Rev. Int. Bot. Appl. Agric. Trop. 24: 25. 1944. – Type not designated.

Pinus kochiana Klotzsch ex K. Koch in Linnaea 22: 296. Jul 1849. – Type not designated.

Pinus koraiensis Siebold & Zucc., Fl. Jap. 2: 28. 1842. – Lectotype (designated here): Siebold & Zuccarini, Fl. Jap. 2: t. 116, f. 5-6. 1842.

Pinus krempfii Lecomte in Bull. Mus. Natl. Hist. Nat. 27: 191. 1921. – Holotype: Vietnam, near Nha Trang, *M. Krempf 1537* (P).

Pinus lagunae (Rob.-Pass.) Passini in Phytologia 63: 337. 2 Oct 1987. – Basionym and type: see under *Pinus cembroides* var. *lagunae*.

Pinus lambertiana Douglas in Trans. Linn. Soc. London 16: 500. ante 19 Mar 1827. – Type not designated.

Pinus latteri Mason in J. Asiat. Soc. Bengal 18: 74. 1849. – Type not designated.

Pinus lawsonii Roezl ex Gordon & Glend., Pinetum, Suppl.: 64. 1862. – Type not designated.

Pinus leiophylla Schiede ex Schltdl. & Cham. in Linnaea 6: 354. Jul-Dec 1831. – Type not designated.

Pinus leiophylla subsp. ***chihuahuana*** (Engelm.) E. Murray in Kalmia 12: 23. Jul 1982. – Basionym and type: see under *Pinus chihuahuana*.

Pinus leiophylla var. ***chihuahuana*** (Engelm.) Shaw [Pines Mexico] in Publ. Arnold Arbor. 1: 14. Mar 1909. – Basionym and type: see under *Pinus chihuahuana*.

Pinus leucodermis Antoine in Österr. Bot. Z. 14: 366. Dec 1864. – Type not designated.

Pinus longaeva D. K. Bailey in Ann. Missouri Bot. Gard. 57: 243. 16 Feb 1971. – Holotype: U.S.A., Nevada, Wheeler Peak Scenic Area, *D. K. Bailey & J. E. Whitson 7001* (COLO; isot.: CAS, NY, RM, US, W).

Pinus luchuensis Mayr in Bot. Centralbl. 58: 149. 24 Jan 1894. – Type not designated.

Pinus lumholtzii B. L. Rob. & Fernald in Proc. Amer. Acad. Arts 30: 122. 27 Aug 1894. – Type not designated.

Pinus lumholtzii var. *microphylla* Carvajal in Phytologia 59: 135. 11 Jan 1986. – Holotype: Mexico, Jalisco, *S. Carvajal 4031* (CREG).

Pinus maëstrensis Bisse in Ciencias (Havana), ser. 10, 2: 2. 1975. – Holotype: Cuba, Sierra Maestra, *O. Oliva & S. Herrera 13904* (HAJB; isot.: JE).

Pinus martinezii E. Larsen in Madroño 17: 217. 23 Jul 1964. – Holotype: Mexico, Michoacan, S. of Paracho, 10 Jan 1960, *H. V. Hinds & E. Larsen 65* (FHO; isot.: CANB, MEXU, NZFRI, UC).

Pinus massoniana Lamb., Descr. Pinus 1: 17, t. 12. Sep 1803. – Lectotype (designated here): Lambert, Descr. Pinus 1: t. 12. Sep 1803.

Pinus massoniana var. *hainanensis* W. C. Cheng & L. K. Fu in Acta Phytotax. Sin. 13(4): 85. Oct 1975. – Holotype: China, Hainan, *Wang Chin 3117* (PE).

Pinus massoniana var. *henryi* (Mast.) C. L. Wu in Acta Phytotax. Sin. 5: 153. 1956. – Basionym and type: see under *Pinus henryi*.

Pinus maximartinezii Rzed. in Ciencia (Mexico) 23: 17, t. 2. 1964. – Holotype: Mexico, Zacatecas, Cerro de Piñones, *J. Rzedowski 18258* (MEXU; isot.: GH, K, P).

Pinus maximinoi H. E. Moore in Baileya 14: 8. 15 Jun 1966. – Holotype: Mexico, *C. T. Hartweg 620* (K; isot.: CGE).

Pinus merkusii Jungh. & de Vriese in Vriese, Pl. Nov. Ind. Bat.: 5, t. 2. 26-31 Mai 1845. – Type not designated.

Pinus merkusii var. *latteri* (Mason) Silba in Phytologia 68: 53. Jan 1990. – Basionym and type: see under *Pinus latteri*.

Pinus michoacana Martínez in Anales Inst. Biol. Univ. Nac. México 15: 1. 1944. – Holotype: Mexico, Michoacan, *M. Martínez 3443* (MEXU).

Pinus michoacana var. *cornuta* Martínez in Anales Inst. Biol. Univ. Nac. México 15: 1. 1944. – Holotype: Mexico, Jalisco, Oct 1941, *M. Martínez 3446* (MEXU).

Pinus michoacana var. *quevedoi* Martínez in Anales Inst. Biol. Univ. Nac. México 15: 1. 1944. – Type not designated.

Pinus monophylla Torr. & Frém. in Frémont, Rep. Exped. Rocky Mts.: 319, t. 4. Mar 1845. – Holotype: U.S.A., California, *J. C. Frémont 367* (NY).

Pinus monophylla var. ***californiarum*** (D. K. Bailey) Silba in Phytologia 68: 54. Jan 1990. – Basionym and type: see under *Pinus californiarum.*

Pinus monophylla var. ***fallax*** (Little) Silba in Phytologia 68: 54. 1990. – Basionym and type: see under *Pinus edulis* var. *fallax.*

Pinus montezumae Lamb., Descr. Pinus, ed. 8°, 1: 39, t. 22. 1832. – Type not designated.

Pinus montezumae var. ***lindleyi*** Loudon, Encycl. Trees Shrubs: 1004, f. 1882-1883. 1842. – Type not designated.

Pinus monticola Douglas ex D. Don in Lambert, Descr. Pinus, ed. 8°, 2: p. s.n. inter 144 et 145. 1832. – Type not designated.

Pinus morrisonicola Hayata in Gard. Chron., ser. 3, 43: 194. 28 Mar 1908. – Holotype: Taiwan, Shohakulin, 21 Jan 1898, *C. Owatari s.n.* (TI).

Pinus mugo Turra in Giorn. Italia Sci. Nat. 1: 152. 1764. – Neotype (designated here): Italy, Monte Baldo, 20 Jul 1890, *Saint-Lager s.n.* (G).

Pinus mugo nothosubsp. ***rotundata*** (Link) Janch. & H. Neumayer in Österr. Bot. Z. 91: 214. 28 Dec 1942. – Basionym and type: see under *Pinus rotundata.*

Pinus mugo subsp. ***uncinata*** (Ramond ex DC.) Domin in Preslia 13-15: 13. 1935. – Basionym and type: see under *Pinus uncinata.*

Pinus muricata D. Don in Trans. Linn. Soc. London 17: 441. 21 Jun - 9 Jul 1836. – Type not designated.

Pinus muricata var. ***borealis*** Axelrod in Univ. Calif. Publ. Geol. Sci. 127: 76. 1983. – Holotype: U.S.A., California, Sonoma Co., *D. I. Axelrod E-3* (GH; isot.: CAS).

Pinus muricata var. ***stantonii*** Axelrod in Univ. Calif. Publ. Geol. Sci. 127: 77. 1983. – Holotype: U.S.A., California, Santa Cruz Island, *H. L. Mason 4096* (UC).

Pinus nelsonii Shaw in Gard. Chron., ser. 3, 36: 122. 20 Aug 1904. – Holotype: Mexico, Miquihuana, near borderline between Tamaulipas and Nuevo León, *E. W. Nelson 4501* (US).

Pinus nigra J. F. Arnold, Reise Mariazell: 8, t. s.n. 1785. – Lectotype (designated here): Arnold, Reise Mariazell: t. post p. 8. 1785.

Pinus nigra subsp. ***dalmatica*** (Vis.) Franco, Dendrol. Florest.: 55. 1943. – Basionym: *Pinus dalmatica* Vis., Fl. Dalmat. 1: 199. 31 Aug - 3 Sep 1842. – Type not designated.

Pinus nigra subsp. ***laricio*** Maire in Bull. Soc. Hist. Nat. Afrique N. 19: 66. 15 Jan 1928. – Type not designated.

Pinus nigra subsp. ***mauretanica*** (Maire & Peyerimh.) Heywood in Feddes Repert. Spec. Nov. Regni Veg. 66: 150. 1 Sep 1962. – Basionym: *Pinus nigra* var. *mauretanica* Maire & Peyerimh. in

Compt. Rend. Hebd. Séances Acad. Sci. 184: 1515. 1927. – Type not designated.

Pinus nigra subsp. ***pallasiana*** (Lamb.) Holmboe [Stud. Veg. Cyprus] in Bergens Mus. Skr., ser. 2, 1(2): 29. Nov-Dec 1914. – Basionym and type: see under *Pinus nigra* var. *pallasiana*.

Pinus nigra subsp. ***salzmannii*** (Dunal) Franco, Dendrol. Florest.: 56. 1943. – Basionym: *Pinus salzmannii* Dunal in Mém. Sect. Sci. Acad. Sci. Montpellier 2: 82. 1851. – Type not designated.

Pinus nigra var. ***caramanica*** (Loudon) Rehder, Man. Cult. Trees: 61. Jan 1927. – Basionym: *Pinus laricio* var. *caramanica* Loudon, Arb. Frutic. Brit.: 2201. 1 Jul 1838. – Type not designated.

Pinus nigra var. ***cebennensis*** (Gren. & Godr.) Rehder in J. Arnold Arbor. 3: 208. Dec 1922. – Basionym: *Pinus laricio* var. *cebennensis* Gren. & Godr., Fl. France 3: 153. Jan-Jun 1855. – Type not designated.

Pinus nigra var. ***pallasiana*** (Lamb.) Asch. & Graebn., Syn. Mitteleur. Fl. 1: 214. 15 Jun 1897. – Basionym: *Pinus pallasiana* Lamb., Descr. Pinus 2: 1, t. 1. 1824. – Type not designated.

Pinus nigra var. ***maritima*** (Aiton) Melville in Kew Bull. 1958: 534. 11 Mai 1959. – Basionym: *Pinus sylvestris* var. *maritima* Aiton, Hort. Kew. 3: 366. 7 Aug - 1 Oct 1789. – Type not designated.

Pinus nubicola J. P. Perry in J. Arnold Arbor. 68: 447. 9 Oct 1987. – Holotype: Guatemala, 40 km E. of San José Pinula, *J. P. Perry GUA.32-79* (GH; isot.: E, K, MEXU).

Pinus oaxacana Mirov in Madroño 14: 145. 27 Jan 1958. – Holotype: Mexico, Oaxaca, *E. W. Nelson 985* (US).

Pinus occidentalis Sw., Prodr.: 103. 20 Jun - 29 Jul 1788. – Lectotype (designated here): "Hispaniola: aux Pins, quartier des Nippes, du côté septentrional", *O. P. Swartz* (BM).

Pinus occidentalis var. ***cubensis*** (Griseb.) Silba in Phytologia Mem. 7: 55. 1984. – Basionym and type: see under *Pinus cubensis*.

Pinus occidentalis var. ***maëstrensis*** (Bisse) Silba in Phytologia 68: 57. Jan 1990. – Basionym and type: see under *Pinus maëstrensis*.

Pinus oocarpa Schiede ex Schltdl. in Linnaea 12: 491. Apr-Sep 1838. – Type not designated.

Pinus oocarpa var. ***manzanoi*** Martínez in Anales Inst. Biol. Univ. Nac. México 11: 70. 1940. – Type not designated.

Pinus oocarpa var. ***microphylla*** Shaw [Pines Mexico] in Publ. Arnold Arbor. 1: 27, t. 20, f. 2, 5, 8-11. Mar 1909. – Lectotype (Styles & McVaugh, 1990): Mexico, Zacatecas, Cerca de Colomas, *Rose 1755* (US; isolect.: A).

Pinus oocarpa var. ***ochoterenai*** Martínez in Anales Inst. Biol.

Univ. Nac. México 11: 65. 1940. – Type not designated.

Pinus orizabensis (D. K. Bailey) D. K. Bailey & Hawksw. in Novon 2: 306. Dec 1992. – Basionym and type: see under *Pinus cembroides* subsp. *orizabensis.*

Pinus palustris Mill., Gard. Dict., ed. 8: *Pinus* No. 14. 16 Apr 1768. – Type not designated.

Pinus parviflora Siebold & Zucc., Fl. Jap. 2: 27, t. 115. 1842. – Lectotype (designated here): "Pinus cembra", Japan, *P. F. von Siebold* (L No. 901.138-394).

Pinus parviflora var. *pentaphylla* (Mayr) A. Henry in Elwes & Henry, Trees Great Britain: 1033. Aug 1910. – Basionym: *Pinus pentaphylla* Mayr, Monogr. Abietin. Japan. Reich.: 78. Nov-Dec 1890. – Type not designated.

Pinus patula Schiede ex Schltdl. & Cham. in Linnaea 6: 354. Jul-Dec 1831. – Type not designated.

Pinus patula subsp. *tecunumanii* (Eguiluz & J. P. Perry) Styles, F.A.O. Forest Genet. Resources Inform. 13: 50. 1984. – Basionym and type: see under *Pinus tecunumanii.*

Pinus patula var. *longipedunculata* Loock ex Martínez, Pinos Mexic., ed. 2: 333, f. 276-280. 1948. – Holotype: Mexico, Oaxaca, *E. E. M. Loock 113* (PRF; isot.: MEXU).

Pinus peuce Griseb., Spic. Fl. Rumel. 2: 349. Jan 1846. – Type not designated.

Pinus pinaster Aiton, Hort. Kew. 3: 367. 7 Aug - 1 Oct 1789. – Type not designated.

Pinus pinaster subsp. *hamiltonii* (Ten.) Villar in Bol. Soc. Esp. Hist. Nat. 33: 427. 1934. – Basionym: *Pinus hamiltonii* Ten., Cat. Orto Bot. Napoli: 90. 1845. – Type not designated.

Pinus pinaster subsp. *renoui* (Villar) Maire [Fl. Afrique N. 1] in Encycl. Biol. 33: 145. 8 Jan 1952. – Basionym: *Pinus pinaster* var. *renoui* Villar in Vol. Jubil. Soc. Sci. Nat. Maroc: 241. 1948. – Type not designated.

Pinus pinceana Gordon & Glend., Pinetum: 204. Jun-Dec 1858. – Type not designated.

Pinus pinea L., Sp. Pl.: 1000. 1 Mai 1753. – Lectotype (designated here by A. Farjon & C. E. Jarvis): illustration in Bauhin & Cherler, Hist. Pl.: 248. 1650.

Pinus pityusa Steven in Bull. Soc. Imp. Naturalistes Moscou 11: 49. 1838. – Type not designated.

Pinus ponderosa Douglas ex Lawson, Agric. Man.: 354. 1836. – Type not designated.

Pinus ponderosa subsp. *scopulorum* (Engelm.) E. Murray in Kalmia 12: 23. Jul 1982. – Basionym and type: see under *Pinus ponderosa* var. *scopulorum.*

Pinus ponderosa subsp. *washoënsis* (Mason & Stockw.) E. Murray in Kalmia 12: 23. Jul 1982. – Basionym and type: see under *Pinus washoënsis.*

Pinus ponderosa var. *arizonica* (Engelm.) Shaw [Pines Mexico] in Publ. Arnold Arbor. 1: 24, t. 4, t. 17, f. 4. Mar 1909. – Basionym and type: see under *Pinus arizonica.*

Pinus ponderosa var. *scopulorum* Engelm. in Watson, Bot. California 2: 126. Jul-Dec 1879. – Type not designated.

Pinus praetermissa Styles & McVaugh in Contr. Univ. Michigan Herb. 17: 310, f. 2. 27 Apr 1990. – Holotype: Mexico, Zacatecas, Nayarit, *Stead & Styles 475* (FHO; isot.: ENCB, MEXU).

Pinus pringlei Shaw in Sargent, Trees & Shrubs 1: 211, t. 100. 8 Apr 1905. – Type not designated.

Pinus pseudostrobus Lindl., Edwards's Bot. Reg. 25: 63. Aug 1839 [& in Allg. Gartenzeitung 7: 325. 1839]. – Type not designated.

Pinus pseudostrobus subsp. *apulcensis* (Lindl.) Stead in Bot. J. Linn. Soc. 89: 269. 20 Jan 1985. – Basionym and type: see under *Pinus apulcensis.*

Pinus pseudostrobus var. *apulcensis* (Lindl.) Shaw [Pines Mexico] in Publ. Arnold Arbor. 1: 19. Mar 1909. – Basionym and type: see under *Pinus apulcensis.*

Pinus pseudostrobus var. *coatepecensis* Martínez, Pinos Mexic., ed. 2: 194. 1948. – Type not designated.

Pinus pseudostrobus var. *estevezii* Martínez in Anales Inst. Biol. Univ. Nac. México 16: 196. 1945.

– Holotype: Mexico, Nuevo León, Santa Catarina, *M. Martínez 3433* (MEXU).

Pinus pseudostrobus var. *oaxacana* (Mirov) S. G. Harrison in Taxon 14: 247. 22 Sep 1965. – Basionym and type: see under *Pinus oaxacana.*

Pinus pumila (Pall.) Regel in Kuester & al., Index Sem. Hort. Bot. Imp. Petrop. 1858: 23. Mar 1859. – Basionym: *Pinus cembra* var. *pumila* Pall., Fl. Ross. 1(1): 5, t. 2. Oct-Dec 1784. – Type not designated.

Pinus pungens Lamb. in Ann. Bot. (König & Sims) 2: 198. Jun 1805. – Type not designated.

Pinus ×quadrifolia Parl. ex Sudw. [Nomencl. Arb. Fl. U.S.] in U.S.D.A. Div. Forest. Bull. 14: 17. 21 Jan 1897 (pro sp.). – Holotype: U.S.A., California, *C. C. Parry 1390* (US No. 60884).

Pinus radiata D. Don in Trans. Linn. Soc. London 17: 442. 21 Jun - 9 Jul 1836. – Type not designated.

Pinus radiata var. *binata* (Engelm.) Lemmon, Cone-Bear. Trees Pacif. Slope, ed. 3: 42. Jul 1895. – Basionym: *Pinus insignis* var. *binata* Engelm. in Watson, Bot. California 2: 127-128. Jul-Dec 1880. – Holotype: Mexico, Guadalupe Island, *E. J. Palmer s.n.* (MO).

Pinus radiata var. *cedrosensis* (J. T. Howell) Silba in Phytologia 68: 60. Jan 1990. – Basionym: *Pinus muricata* var. *cedrosensis* J. T. Howell in Leafl. W. Bot. 3: 7. 4

Feb 1941. – Holotype: Mexico, Baja California, Isla Cedros, *H. L. Mason 2030* (CAS).

Pinus remota (Little) D. K. Bailey & Hawksw. in Phytologia 44: 129. 13 Sep 1979. – Basionym and type: see under *Pinus cembroides* var. *remota.*

Pinus resinosa Aiton, Hort. Kew. 3: 367. 7 Aug - 1 Oct 1789. – Type not designated.

Pinus ×rhaetica Brügger in Flora 47: 150. 16 Mar 1864 (pro sp.). – Neotype (Christensen, 1987): Switzerland, Upper Engadin, 27 Jun 1980, *K. I. Christensen CH2-5a* (Z; isoneot.: C).

Pinus rigida Mill., Gard. Dict., ed. 8: *Pinus* No. 10. 16 Apr 1768. – Type not designated.

Pinus rigida subsp. ***serotina*** (Michx.) R. T. Clausen in Torreya 39: 126. 1939. – Basionym and type: see under *Pinus serotina.*

Pinus rigida var. ***serotina*** (Michx.) Hoopes, Book Evergr.: 120. Jan-Aug 1868. – Basionym and type: see under *Pinus serotina.*

Pinus ×rotundata Link in Flora 10: 218. 14 Apr 1827 [& in Abh. Königl. Akad. Wiss. Berlin 1827: 168. 1830], pro sp. – Neotype (Christensen, 1987): [locality not mentioned], 2/6 Jul 1981, *K. I. Christensen A4* (B; isoneot.: C).

Pinus roxburghii Sarg., Silva 11: 19. 29 Apr 1898. – Type not designated.

Pinus rudis Endl., Syn. Conif.: 151. Mai-Jun 1847. – Type not designated.

Pinus rzedowskii Madrigal & M. Caball. in Bol. Técn. Secr. Agric. Ganad. Subsecr. Forest. Fauna 26: 1. 1969. – Holotype: Mexico, Michoacan, *S. X. Madrigal 2202* (INIF; isot.: K, P, US).

Pinus sabiniana Douglas ex D. Don in Lambert, Descr. Pinus, ed. 8°, 2: p. s.n. inter 144 et 145, t. 80. 1832. – Type not designated.

Pinus serotina Michx., Fl. Bor.-Amer. 2: 205. 19 Mar 1803. – Type not designated.

Pinus sibirica Du Tour in Déterville, Nouv. Dict. Hist. Nat. 18: 18. 23 Oct 1803. – Type not designated.

Pinus squamata X. W. Li in Acta Bot. Yunnan. 14: 259. 1992. – Holotype: China, Yunnan, *X. W. Li 91250* (SWCF).

Pinus strobiformis Engelm. in Wislizenus, Mem. Tour N. Mexico: 102. Apr 1848. – Holotype: Mexico, Chihuahua, *F. A. Wislizenus 155* (MO).

Pinus strobus L., Sp. Pl.: 1001. 1 Mai 1753. – Lectotype (designated here by A. Farjon & C. E. Jarvis): "5 K Strobus", *P. Kalm s.n.* in Herb. Linn. No. 1135.10 (LINN).

Pinus strobus var. ***chiapensis*** Martínez in Anales Inst. Biol. Univ. Nac. México 11: 81. 1940. – Lectotype (Andresen, 1964): Mexico, Chiapas, *M. Martínez s.n.* (MEXU No. 97).

Pinus sylvestris L., Sp. Pl.: 1000. 1 Mai 1753. – Lectotype (Farjon & Jarvis in Jarvis & al., 1993):

illustration of "Pinus sylvestris" in Daléchamps, Hist. General. Pl.: 45. 1586.

Pinus sylvestris subsp. ***hamata*** (Steven) Fomin, Věstn. Tiflissk. Bot. Sada 34: 20. 1914. – Basionym and type: see under *Pinus sylvestris* var. *hamata*.

Pinus sylvestris var. ***hamata*** Steven in Bull. Soc. Imp. Naturalistes Moscou 11: 52. 1838. – Lectotype (Christensen, 1987): U.S.S.R., Caucasus Mts., *H. Wittmann s.n.* (H No. 1002536, right hand specimen).

Pinus sylvestris var. ***lapponica*** Hartm., Handb. Skand. Fl., ed. 5: 214. Dec 1849 - Jan 1850. – Neotype (Christensen, 1987): [locality not mentioned], 27 Jul 1863, *E. M. Fries s.n.* (UPS).

Pinus sylvestris var. ***mongolica*** Litv., Sched. Herb. Fl. Ross. 5: 160. Dec 1905 [& in Repert. Spec. Nov. Regni Veg. 4: 11. 1 Mai 1907]. – Holotype: Mongolia, along railroad near "Charchonte", 21 Jun 1902, *D. Litvinov s.n.* (LE).

Pinus sylvestris var. ***sibirica*** Ledeb., Fl. Altaica 4: 199. Jul-Dec 1833. – Type not designated.

Pinus sylvestris var. ***sylvestriformis*** (Taken.) W. C. Cheng & C. D. Chu in Fl. Reipubl. Pop. Sin. 7: 246. Dec 1978. – Basionym: *Pinus densiflora* f. *sylvestriformis* Taken. in J. Jap. Forest. Soc. 24: 120, f. 1. 1942. – Type not designated.

Pinus tabuliformis Carrière, Traité Gén. Conif., ed. 2: 510. 15 Jan 1867. – Type not designated.

Pinus tabuliformis var. ***densata*** (Mast.) Rehder in J. Arnold Arbor. 7: 23. 4 Mar 1926. – Basionym and type: see under *Pinus densata.*

Pinus tabuliformis var. ***henryi*** (Mast.) C. T. Kuan, Fl. Sichuan. 2: 113. 1 Jan 1983. – Basionym and type: see under *Pinus henryi.*

Pinus tabuliformis var. ***mukdensis*** (Nakai) Uyeki in J. Chosen Nat. Hist. Soc. 3: 35. 1925. – Basionym: *Pinus mukdensis* Nakai in Bot. Mag. (Tokyo) 33: 195. 1919. – Holotype: Korea, *H. Uyeki 2350* (TI).

Pinus taeda L., Sp. Pl.: 1000. 1 Mai 1753. – Lectotype (designated here by A. Farjon & C. E. Jarvis): U.S.A., *J. Clayton 496* (BM).

Pinus taiwanensis Hayata in J. Coll. Sci. Imp. Univ. Tokyo 30(1): 307. 13 Feb 1911. – Lectotype (designated here): Taiwan, Central Mts., 27 Nov 1906, *T. Kawakami & U. Mori 2097* (TI).

Pinus taiwanensis var. ***damingshanensis*** W. C. Cheng & L. K. Fu in Acta Phytotax. Sin. 13(4): 85. Oct 1975. – Holotype: China, *Damingshan exploration group 74297* (PE).

Pinus tecunumanii Eguiluz & J. P. Perry in Revista Ci. Forest. 8: 4. 1983. – Holotype: Guatemala, San Jéronimo, *T. Eguiluz/INAFOR 2* (A No. 3786).

Pinus teocote Schiede ex Schltdl. & Cham. in Linnaea 5: 76. Jan 1830. – Holotype: Mexico, near Pico de Orizaba, *C. J. W. Schiede & F. Deppe s.n.* (HAL).

Pinus thunbergii Parl. in Candolle, Prodr. 16(2): 388. med Jul 1868. – Type not designated.

Pinus torreyana Parry ex Carrière, Traité Gén. Conif.: 326. Jan 1855. – Neotype (Haller, 1986): U.S.A., California, near Soledad Valley, *J. R. Haller 10450* (UCSB).

Pinus torreyana subsp. *insularis* J. R. Haller in Syst. Bot. 11: 45. 28 Feb 1986. – Holotype: U.S.A., California, Santa Rosa Island, *J. R. Haller 10448* (UCSB; isot.: NY, US).

Pinus tropicalis Morelet in Rev. Hort. Côte d'Or 1: 106. 1851. – Type not designated.

Pinus uncinata Ramond ex DC. in Lamarck & Candolle, Fl. Franç., ed. 3, 3: 726. 17 Sep 1805. – Neotype (Christensen, 1987): [locality not mentioned], 5/12 Jul 1980, *K. I. Christensen F6* (G).

Pinus virginiana Mill., Gard. Dict., ed. 8: *Pinus* No. 9. 16 Apr 1768. – Type not designated.

Pinus wallichiana A. B. Jacks. in Bull. Misc. Inform., Kew 1938: 85. 23 Mar 1938. – Type not designated.

Pinus wangii Hu & W. C. Cheng in Bull. Fan Mem. Inst. Biol., ser. 2, 1: 191. 1948. – Holotype: China, S.E. Yunnan, *C. W. Wang 85830* (KUN).

Pinus washoënsis Mason & Stockw. in Madroño 8: 62. 31 Mar 1945. – Holotype: U.S.A., Nevada, Mt. Rose, *H. L. Mason 12370* (UC; isot.: CAS, NY).

Pinus wincesteriana Gordon in J. Hort. Soc. London 2: 158. 1847. – Type not designated.

Pinus yunnanensis Franch. in J. Bot. (Morot) 13: 253. Aug 1899. – Holotype: China, Yunnan, *P. J. M. Delavay 569* (P).

Pinus yunnanensis var. *pygmaea* (J. R. Xue) J. R. Xue in Fl. Reipubl. Pop. Sin. 7: 258. Dec 1978. – Basionym: *Pinus densata* var. *pygmaea* J. R. Xue in Acta Phytotax. Sin. 13(4): 85. Oct 1975. – Holotype: China, Yunnan, *Y. C. Hsüeh 156* (PE).

Pinus yunnanensis var. *tenuifolia* W. C. Cheng & Y. W. Law in Acta Phytotax. Sin. 13(4): 85. Oct 1975. – Holotype: China, Yunnan, Chinese collector No.1038 (PE).

Pseudolarix Gordon, *nom. cons.*

Pseudolarix amabilis (J. Nelson) Rehder in J. Arnold Arbor. 1: 53. 21 Jul 1919. – Basionym: *Larix amabilis* J. Nelson, Pinaceae: 84. Jan-Mai 1866. – Type not designated.

Pseudotsuga Carrière

Pseudotsuga sect. *Sina* E. Murray in Kalmia 14: 16. Jan 1984. – Holotype: *Pseudotsuga sinensis* Dode.

Pseudotsuga brevifolia W. C. Cheng & L. K. Fu in Acta Phytotax. Sin. 13(4): 83, t. 16. Oct 1975. – Holotype: China, Guangxi, near Longzhou, Chinese collector No. 40103 (PE).

Pseudotsuga forrestii Craib in Notes Roy. Bot. Gard. Edinburgh 11: 189. Nov 1919. – Lectotype (Farjon, 1988): China, Yunnan, Lancang (Mekong) Valley, *G. Forrest 13003* (E).

Pseudotsuga gaussenii Flous in Bull. Soc. Hist. Nat. Toulouse 69: 417, f. 1-11. 1936. – Holotype: China, Zhejiang, *R. C. Ching 2144* (NY; isot.: P).

Pseudotsuga glauca (Beissn.) Mayr in Mitt. Deutsch. Dendrol. Ges. 10: 57. 1901. – Basionym: *Tsuga douglasii* var. *glauca* Beissn. in Jäger & Beissner, Zier-geh. Gärt. Park., ed. 2: 446. 1884. – Type not designated.

Pseudotsuga japonica (Shiras.) Beissn. in Mitt. Deutsch. Dendrol. Ges. 5: 62. 1896. – Basionym: *Tsuga japonica* Shiras. in Bot. Mag. (Tokyo) 9: 84, t. 3. 1895. – Type not designated.

Pseudotsuga macrocarpa (Vasey) Mayr, Wald. Nordamer.: 278, t. 6, 8, 9. Sep-Nov 1889. – Basionym: *Abies macrocarpa* Vasey in Gard. Monthly & Hort. 18: 21. 1876. – Type not designated.

Pseudotsuga macrolepis Flous in Bull. Soc. Hist. Nat. Toulouse 66: 219. 1934. – Holotype: Mexico, near Moran, *C. T. Hartweg 439* (P; isot.: NY).

Pseudotsuga menziesii (Mirb.) Franco, Conif. Duarum Nom.: 4. 1950 [& in Bol. Soc. Brot., ser. 2, 24: 74. 1950]. – Basionym: *Abies menziesii* Mirb. in Mém. Mus. Hist. Nat. 13: 63, 70. 1825. – Type not designated.

Pseudotsuga menziesii var. *glauca* (Beissn.) Franco, Conif. Duarum Nom.: 6. 1950 [& in Bol. Soc. Brot., ser. 2, 24: 76. 1950]. – Basionym and type: see under *Pseudotsuga glauca*.

Pseudotsuga sinensis Dode in Bull. Soc. Dendrol. France 23-24: 58. 1912. – Holotype: China, Yunnan, 1911, *E. E. Maire s.n.* (P).

Pseudotsuga sinensis var. *brevifolia* (W. C. Cheng & L. K. Fu) Farjon & Silba in Phytologia 68: 71. Jan 1990 [& Farjon in Regnum Veg. 121: 187. Nov 1990]. – Basionym and type: see under *Pseudotsuga brevifolia*.

Pseudotsuga sinensis var. *forrestii* (Craib) Silba in Phytologia 68: 72. Jan 1990. – Basionym and type: see under *Pseudotsuga forrestii*.

Pseudotsuga sinensis var. *gaussenii* (Flous) Silba in Phytologia 68: 71. Jan 1990 [& Farjon in Regnum Veg. 121: 185. Nov 1990]. – Basionym and type: see under *Pseudotsuga gaussenii*.

Pseudotsuga wilsoniana Hayata, Icon. Pl. Formos. 5: 204, t. 15. 25 Nov 1915. – Holotype: Taiwan, Mt. Morrison, Dec 1908, *U. Mori s.n.* (TI, cone).

Tsuga (Endl.) Carrière

Tsuga subg. ***Hesperopeuce*** (Engelm.) Ueno in J. Inst. Polytechn. Osaka City Univ., ser. D, Biol. 8: 194. 1957. – Basionym and type: see under *Tsuga* sect. *Hesperopeuce*.

Tsuga sect. ***Hesperopeuce*** Engelm. in Watson, Bot. California 2: 121. Jul-Dec 1880. – Holotype: *Tsuga pattoniana* (A. Murray bis) Engelm. [= *Tsuga mertensiana* (Bong.) Carrière].

Tsuga canadensis (L.) Carrière, Traité Gén. Conif.: 189. Jan 1855. – Basionym: *Pinus canadensis* L., Sp. Pl., ed. 2: 1421. Jul-Aug 1763. – Lectotype (Fernald & Weatherby, 1932): E. North America, *J. Clayton 547* (BM).

Tsuga caroliniana Engelm. in Bot. Gaz. 6: 223. 1881. – Lectotype (designated here): U.S.A., South Carolina, 1879, *A. H. Curtiss s.n.* ex herb. Engelm. (MO).

Tsuga chinensis (Franch.) Pritz. in Bot. Jahrb. Syst. 29: 217. 4 Sep 1900. – Basionym: *Abies chinensis* Franch. in J. Bot. (Morot) 13: 259. Aug 1899. – Holotype: China, Sichuan, *P. G. Farges 808* (P).

Tsuga chinensis var. ***formosana*** (Hayata) H. L. Li & H. Keng, Taiwania 5: 64. Oct 1954. – Basionym and type: see under *Tsuga formosana*.

Tsuga chinensis var. ***oblongisquamata*** W. C. Cheng & L. K. Fu in Acta Phytotax. Sin. 13(4): 83. Oct 1975. – Holotype: China, W.

Hubei, Chinese collector No. 950 (PE).

Tsuga chinensis var. ***robusta*** W. C. Cheng & L. K. Fu in Acta Phytotax. Sin. 13(4): 83. Oct 1975. – Holotype: China, Hubei, Yangtse Gorges, *Chen & al. 2000* (PE).

Tsuga diversifolia (Maxim.) Mast. in J. Linn. Soc., Bot. 18: 514. 1881. – Basionym: *Abies diversifolia* Maxim. [Diagn. Pl. Nov. Jap. 6] in Bull. Acad. Imp. Sci. Saint-Pétersbourg 12: 229. 13 Nov 1867. – Type not designated.

Tsuga dumosa (D. Don) Eichler in Engler & Prantl, Nat. Pflanzenfam. 2(1): 80. Jun 1887. – Basionym: *Pinus dumosa* D. Don, Prodr. Fl. Nepal.: 55. 26 Jan - 1 Feb 1825. – Type not designated.

Tsuga formosana Hayata in Gard. Chron., ser. 3, 43: 194. 28 Mar 1908. – Holotype: Taiwan, Mt. Morrison, 1 Nov 1905, *S. Nagasawa 552* (TI).

Tsuga forrestii Downie in Notes Roy. Bot. Gard. Edinburgh 14: 18. Apr 1923. – Holotype: China, Yunnan, Likiang Range, *G. Forrest 17169* (E).

Tsuga heterophylla (Raf.) Sarg., Silva 12: 73, t. 605. 10 Jan 1899. – Basionym: *Abies heterophylla* Raf. in Atlantic J. 1: 119. aut 1832. – Type not designated.

Tsuga hookeriana (A. Murray bis) Carrière, Traité Gén. Conif., ed. 2: 252. 15 Jan 1867. – Basionym: *Abies hookeriana* A. Murray bis in Edinburgh New Philos. J. 1: 289. 1855. – Type not designated.

Tsuga ×*jeffreyi* (A. Henry) A. Henry in Proc. Roy. Irish Acad., B, 34: 55. 1919. – Basionym: *Tsuga pattoniana* var. *jeffreyi* A. Henry in Elwes & Henry, Trees Great Britain 2: 231. Sep 1907. – Type not designated.

Tsuga longibracteata W. C. Cheng in Contr. Biol. Lab. Sci. Soc. China, Bot. Ser. 7(1): 1. 1 Jan 1932. – Holotype: China, Guizhou, Van-ching Shan, Yingkiang, *Y. Tsiang 7712* (NAS; isot.: E).

Tsuga mertensiana (Bong.) Carrière, Traité Gén. Conif., ed. 2: 250. 15 Jan 1867. – Basionym: *Pinus mertensiana* Bong. in Mém. Acad. Imp. Sci. St.-Pétersbourg, ser. 6, Sci. Math. 2: 163. 1832. – Type not designated.

Tsuga mertensiana subsp. *grandicona* Farjon in Proc. Kon. Ned. Akad. Wetensch., C, Bot. 91: 39. 28 Mar 1988. – Holotype: U.S.A., California, Twin Lakes, *B. Willard 54* (RM).

Tsuga mertensiana var. *jeffreyi* (A. Henry) C. K. Schneid. in Silva-Tarouca & Schneider, Freiland-Nadelhölz., ed. 2: 303. Apr 1923. – Basionym and type: see under *Tsuga* ×*jeffreyi*.

Tsuga sieboldii Carrière, Traité Gén. Conif.: 186. Jan 1855. – Lectotype (designated here): "in Japonia", *P. F. von Siebold* comm. 1842 ex herb. Zuccarini No. 265 (M).

Tsuga yunnanensis (Franch.) Pritz. in Bot. Jahrb. Syst. 29: 217. 4 Sep 1900. – Basionym: *Abies yunnanensis* Franch. in J. Bot. (Morot) 13: 258. Aug 1899. – Type not designated.

References

Andresen, J. W. 1964. The taxonomic status of *Pinus chiapensis*. *Phytologia* 10: 417-421.

Barrett, W. H. G. & Golfari, L. 1962. Descripción de dos nuevas variedades del Pino del Caribe. *Caribbean Forest.* 23: 59-71.

Christensen, K. I. 1987. Taxonomic revision of the *Pinus mugo* complex and *P.* ×*rhaetica (P. mugo* × *sylvestris) (Pinaceae). Nordic J. Bot.* 7: 383-408.

Farjon, A. 1988. Taxonomic notes on *Pinaceae* I. *Proc. Kon. Ned. Akad. Wetensch., C, Bot.* 91: 31-42.

– 1989. A second revision of the genus *Keteleeria* Carrière. (Taxonomic notes on *Pinaceae* II). *Notes Roy. Bot. Gard. Edinburgh* 46: 81-99.

– & Rushforth, K. D. 1989. A classification of *Abies* Miller *(Pinaceae). Notes Roy. Bot. Gard. Edinburgh* 46: 59-79.

Fernald, M. L. & Weatherby, C. A. 1932. *Picea glauca,* forma *parva. Rhodora* 34: 187-189.

Greuter, W. 1991a. Draft lists of NCU: First progress report. *Taxon* 40: 521-524.

– 1991b. (20-41) Proposals to amend the Code, and report of Special Committee 6B: Provisions for granting nomenclatural protection to listed names in current use. *Taxon* 40: 669-677.

– , Burdet, H. M., Chaloner, W. G., Demoulin, V., Grolle, R., Hawksworth, D. L., Nicolson, D. H., Silva, P. C., Stafleu, F. A., Voss, E. G. & McNeill, J. (ed.) 1988. International code of botanical nomenclature adopted by the Fourteenth International Botanical Congress, Berlin, July-August 1987. *Regnum Veg.* 118.

Haller, J. R. 1986. Taxonomy and relationships of the mainland and island populations of *Pinus torreyana (Pinaceae). Syst. Bot.* 11: 39-50.

Hara, H. & Brummitt, R. K. 1980. the problem of *Pseudolarix kaempferi:* type method or circumscription method? *Taxon* 29: 315-317.

Hawksworth, D. L. (ed.) 1991. Improving the stability of names: needs and options. Proceedings of an international symposium, Kew, 20-23 February 1991. *Regnum Veg.* 123.

– & Greuter, W. 1989. Report of the first meeting of a working group on lists of names in current use. *Taxon* 38: 142-148.

Holmgren, P. K., Holmgren, N. H. & Barnett, L. C. (ed.) 1990. Index herbariorum. Part I: the herbaria of the world. Eighth edition. *Regnum Veg.* 120.

Hunt, D. R. 1991. Stabilization of names in succulent plants. *Regnum Veg.* 123: 151-155.

Jarvis, C. E., Barrie, F. R., Allan, D. M. & Reveal, J. L. 1993. A list of Linnaean generic names and their types. *Regnum Veg.* 127.

Lambert, A. B. 1837. *A description of the genus* Pinus, ed. 2, 3. London.

Styles, B. T. & McVaugh, R. 1990. A mexican pine promoted to specific status: *Pinus praetermissa. Contr. Univ. Michigan Herb.* 17: 307-312.

SPECIES NAMES IN CURRENT USE IN THE *LEMNACEAE* (MONOCOTYLEDONES)

By Elias Landolt

This list is largely based on the taxonomic part of my monographic study of the duckweed family (Landolt, 1986). The nomenclature and typification follows the monographic treatment, except for a few technical corrections and one name change *(Lemna minuta* adopted instead of *Lemna minuscula)*, where advice given by J. L. Reveal has been followed.

It has not been thought advisable, for the time being, to extend the list to ranks other than species, since names at such other ranks are little used and of scant taxonomical significance in this family.

Technical explanations on format and abbreviations are given in the general preface to this booklet. When citing type specimens, the country, province and place of origin, the date of collection and the collector(s) with collection number are indicated as fully as possible, and as required for the unambiguous designation of the specimen. Herbaria are designated by the standard abbreviations used in Holmgren & al. (1990).

Lemna L.

Lemna aequinoctialis Welw. [Apont.] in Ann. Cons. Ultramarino 1: 578. Dec 1859. – Lectotype (Landolt, 1986): Angola, prov. Luanda, distr. Luanda, 1858, *F. Welwitsch 206* (STU; isolectot.: BM, G, K, ZT).

Lemna disperma Hegelm. in Bot. Zeitung (Berlin) 29: 654. 29 Sep 1871. – Lectotype (Landolt, 1986): Australia, West Australia, Swan River, *Drummond 1845* (STU; isolectot.: K).

Lemna ecuadoriensis Landolt in Veröff. Geobot. Inst. E.T.H. Stiftung Rübel Zürich 70: 22. Dec 1980. – Holotype: Ecuador, prov. El Oro, between Machala and Santa Rosa, 1974, *T. Plowman, L. Jacobs & E. W. Davis 4609* (U; isot.: F, GH).

Lemna gibba L., Sp. Pl.: 970. 1 Mai 1753. – Type not designated.

Lemna japonica Landolt in Veröff. Geobot. Inst. E.T.H. Stiftung Rübel Zürich 70: 23. Dec 1980. – Holotype: Japan, Kyushu, Fukuoka, Mitumaki-machi, 1969, *K. Okutomi,* ETH No. 7182 (ZT).

Lemna minor L., Sp. Pl.: 970. 1 Mai 1753. – Lectotype (Landolt, 1986): Herb. Linn. No. 1093.2 (LINN).

Lemna minuta Kunth in Humboldt & al., Nov. Gen. Sp., ed. 4°, 1: 372. Aug 1816. – Neotype (Reveal, 1990): Chile, Santiago, Mai

1857, *R. A. Philippi* (STU; isoneot.: BM, G, GOET, K, LE,
MEL, MO, S, SGO, UPS).

Lemna obscura (Austin) Daubs in
Illinois Biol. Monogr. 34: 20. Dec
1965. – Basionym: *Lemna minor*
var. *obscura* Austin in Gray, Man-
ual, ed. 5: 479. Jun-Dec 1867. –
Lectotype (Daubs, 1965): U.S.A.,
New Jersey, New Durham
Meadows, 1862, *C. T. Austin
8050* (MO; isolectot.: NY, STU).

Lemna perpusilla Torr., Fl. New
York 2: 245. 1843. – Lectotype
(Landolt, 1986): U.S.A., New
York, Staten Island, 1829, *J. Tor-
rey* (STU; isolectot.: GH, KANU,
MO, NY).

Lemna tenera Kurz in J. Asiat. Soc.
Bengal 40: 78. 1871. – Lectotype
(Landolt, 1986): Burma, Pegu,
Pazwoodoung Valley, 1870, *S.
Kurz* (STU; isolectot.: K).

Lemna trisulca L., Sp. Pl.: 970. 1
Mai 1753. – Lectotype (Landolt,
1986): Herb. Linn. No. 1093.1
(LINN).

Lemna turionifera Landolt in
Aquatic Bot. 1: 355. 1975. – Holotype: U.S.A., Montana, Lincoln
County, 20 km W. of Davensport,
1953, *W. M. Hiesey 196*, ETH No.
6573 (ZT).

Lemna valdiviana Phil. in Linnaea
33: 239. Aug 1864. – Lectotype
(Landolt, 1986): Chile, Prov. Val-
divia , S. Juan, Trembladerilla,
1861, *R. A. Philippi* (STU; isolectot.: MO).

Spirodela Schleid.

Spirodela intermedia W. Koch in
Ber. Schweiz. Bot. Ges. 41: 114.
Apr 1932. – Holotype: Uruguay,
Montevideo, 1877, *J. Arechava-
letta 2502* (ZT).

Spirodela polyrrhiza (L.) Schleid.
in Linnaea 12: 392. Apr-Sep
1838. – Basionym: *Lemna polyrrhiza* L., Sp. Pl.: 970. 1 Mai
1753. – Type not designated.

Spirodela punctata (G. Mey.) C. H.
Thomps. in Rep. (Annual) Missouri Bot. Gard. 9: 28. 1 Nov
1897. – Basionym: *Lemna punctata* G. Mey., Prim. Fl. Esseq.:
262. Nov 1818. – Neotype
(Thompson, 1897): Chile, Tierra
del Fuego Island, Orange Harbor,
1838, Wilkes Expedition (US;
isoneot.: DS, GH, KANU, MO).

Wolffia Horkel ex Schleid.,
nom. cons.

Wolffia angusta Landolt in Veröff.
Geobot. Inst. E.T.H. Stiftung Rü-
bel Zürich 70: 29. Dec 1980. –
Holotype: Australia, New South
Wales, N.E. of Newcastle, c. 8 km
W. of Seaham, 1970, *B. G. Briggs
& L. A. S. Johnson,* ETH No. 7274
(ZT; isot.: NSW).

Wolffia arrhiza (L.) Horkel ex
Wimm., Fl. Schles., ed. 3: 140.
1857. – Basionym: *Lemna arrhiza*
L., Mant. Pl.: 294. Oct 1771. –
Type not designated.

Wolffia australiana (Benth.) Hartog & Plas in Blumea 20: 151. 22
Sep 1972. – Basionym: *Wolffia
arrhiza* var. *australiana* Benth.,

Fl. Austral. 7: 162. 23-30 Mar 1878. – Lectotype (Hartog & Plas, 1972): Australia, Victoria, Mount Emu Creek, 1874, *F. von Mueller* (K; isolectot.: BRI, FI, MEL, STU).

Wolffia borealis (Engelm.) Landolt in Ber. Geobot. Inst. E.T.H. Stiftung Rübel Zürich 44: 137. 1977. – Basionym: *Wolffia brasiliensis* var. *borealis* Engelm. in Hegelmaier, Lemnac.: 127. Oct-Nov 1868. – Lectotype (Landolt, 1986): U.S.A., Illinois, between Springfield and Athens, 1867, *E. Hall* ex herb. Engelmann (STU; isolectot.: DS, F, G, GH, H, KANU, L, MIN, MO).

Wolffia brasiliensis Wedd. in Ann. Sci. Nat., Bot., ser. 3, 12: 170. Sep 1849. – Lectotype (Landolt, 1986): Brazil, Mato Grosso, 1845, *M. H. A. Weddell* (STU; isolectot.: K, L, MO).

Wolffia columbiana H. Karst. in Bot. Untersuch. (Berlin) 1: 103. 1865. – Lectotype (Landolt, 1986): Columbia, Santa Marta, *H. Karsten* (STU).

Wolffia elongata Landolt in Veröff. Geobot. Inst. E.T.H. Stiftung Rübel Zürich 70: 27. Dec 1980. – Holotype: Columbia, Dept. Atlantico, Barranquilla, 1948, *G. Gutiérrez & F. E. Berkeley 18 C 006* (COL; isot.: S, US).

Wolffia globosa (Roxb.) Hartog & Plas in Blumea 18: 367. 31 Dec 1970. – Basionym: *Lemna globosa* Roxb., Fl. Ind., ed. 1832, 3: 565. Oct-Dec 1832. – Type not designated.

Wolffia microscopica (Griff. ex Voigt) Kurz in J. Linn. Soc., Bot. 9: 265. 1866. – Basionym: *Grantia microscopica* Griff. ex Voigt, Hort. Suburb. Calcutt.: 692. 1845. – Lectotype (Landolt, 1986): India , Calcutta, *W. Griffith* (K).

Wolffiella (Hegelm.) Hegelm.

Wolffiella denticulata (Hegelm.) Hegelm. in Bot. Jahrb. Syst. 21: 305. 6 Aug 1895. – Basionym: *Wolffia denticulata* Hegelm., Lemnac.: 133. Oct-Nov 1868. – Holotype: South Africa, Cape, *Krauss* ex herb. A. Braun (STU; isot.: FI, G, JE, M, MO, TUB, Z).

Wolffiella gladiata (Hegelm.) Hegelm. in Bot. Jahrb. Syst. 21: 304. 6 Aug 1895. – Basionym: *Wolffia gladiata* Hegelm., Lemnac.: 133. Oct-Nov 1868. – Lectotype (Landolt, 1986): Mexico, 1868, *L. Hahn* (STU).

Wolffiella hyalina (Delile) Monod in Mém. Soc. Hist. Nat. Afrique N., Hors Sér. 2: 242. 6 Mai 1949. – Basionym: *Lemna hyalina* Delile, Descr. Egypte, Hist. Nat.: 75. 1813 – Holotype: Egypt, Delile 877 (MPU; isot.: BM, MPU, P).

Wolffiella lingulata (Hegelm.) Hegelm. in Bot. Jahrb. Syst. 21: 303. 6 Aug 1895. – Basionym: *Wolffia lingulata* Hegelm., Lemnac.: 132. Oct-Nov 1868. – Holotype: Mexico, May 1868, L. Hahn (STU).

Wolffiella neotropica Landolt in Veröff. Geobot. Inst. E.T.H. Stiftung Rübel Zürich 70: 26. Dec

1980. – Holotype: Brazil, Rio de Janeiro, Recreiro Bandairantes, Canal das Tachas, 1969, *F. Segadas-Vianna*, ETH No. 7225 (ZT).

Wolffiella oblonga (Phil.) Hegelm. in Bot. Jahrb. Syst. 21: 303. 6 Aug 1895. – Basionym: *Lemna oblonga* Phil. in Linnaea 29: 45. Feb-Mar 1858. – Lectotype (Landolt, 1986): Chile, Santiago de Chile, Mai 1857, *R. A. Philippi* (STU; isolectot.: BM, G, GOET, K, L, LE, MEL, MO, S, SGO, UPS, ZT).

Wolffiella repanda (Hegelm.) Monod in Mém. Soc. Hist. Nat. Afrique N., Hors Sér. 2: 242. 6 Mai 1949. – Basionym: *Wolffia repanda* Hegelm. in J. Bot. 3: 113. Apr 1865. – Holotype: Angola, distr. Luanda, Bemposta, 1854, *F. Welwitsch 205* (STU; isot.: BM, C, G, K, L, P).

Wolffiella rotunda Landolt in Veröff. Geobot. Inst. E.T.H. Stiftung Rübel Zürich 70: 26. Dec 1980. – Holotype: Zimbabwe, distr. Urungwe, Zambesi Valley, W. end of Kariba Gorge, 1953, *H. Wild 4265* (SRGH; isot.: B, K, LISU, MO, PRE, S, SRGH).

Wolffiella welwitschii (Hegelm.) Monod in Mém. Soc. Hist. Nat. Afrique N., Hors Sér. 2: 242. 6 Mai 1949. – Basionym: *Wolffia welwitschii* Hegelm. in J. Bot. 3: 114. Apr 1865. – Holotype: Angola, distr. Ambriz, near Quizembo, 1853, *F. Welwitsch 209* (STU; isot.: BM, C, G, H, K, L).

References

Daubs, E. H. 1965. A monograph of *Lemnaceae. Illinois Biol. Monogr.* 34.

Hartog, C. den & Plas, F. van der 1972. The Australian species of *Wolffia (Lemnaceae). Blumea* 20: 151-153.

Holmgren, P. K., Holmgren, N. H. & Barnett, L. C. 1990. Index herbariorum. Part I. The herbaria of the world, ed. 8. *Regnum Veg.* 120.

Holmgren, P. K., Holmgren, N. H. & Barnett, L. C. (ed.) 1990. Index herbariorum. Part I. The herbaria of the world, ed. 8. *Regnum Veg.* 120.

Landolt, E. 1986. Biosystematic investigations in the family of duckweeds *(Lemnaceae)* (vol. 2). The family of *Lemnaceae* – a monographic study. Volume 1. *Veröff. Geobot. Inst. E.T.H. Stiftung Rübel Zürich* 71.

Reveal, J. L. 1990. The neotypification of *Lemna minuta* Humb., Bonpl. & Kunth, an earlier name for *Lemna minuscula* Herter *(Lemnaceae). Taxon* 39: 328-330.

Thompson, C. H. 1897. A revision off the American *Lemnaceae* occurring north of Mexico. *Rep. (Annual) Missouri Bot. Gard.* 9: 21-42.

TABLE OF CONTENTS

Addresses of the authors:

T. Ahti, Department of Botany,University, Unioninkatu 44, SF-00170 Helsinki, Finland.

Aljos Farjon, Department of Plant Ecology & Evolutionary Biology,
University of Utrecht, Heidelberglaan 2, NL-3584 CS Utrecht, The
Netherlands.

Werner Greuter, Botanischer Garten und Botanisches Museum Berlin-
Dahlem, Königin-Luise-Str. 6-8, D-14191 Berlin, Germany.

Elias Landolt, Geobotanisches Institut der ETH, Stiftung Rübel, Zürich-
bergstr. 38, CH-8044 Zürich, Switzerland.

J. I. Pitt, CSIRO Food Research Laboratory, P. O. Box 52, North Ryde,
N.S.W. 2113, Australia.